INFORMAL GEOMETRY

AMY WANGSNESS WEHE
Fitchburg State University

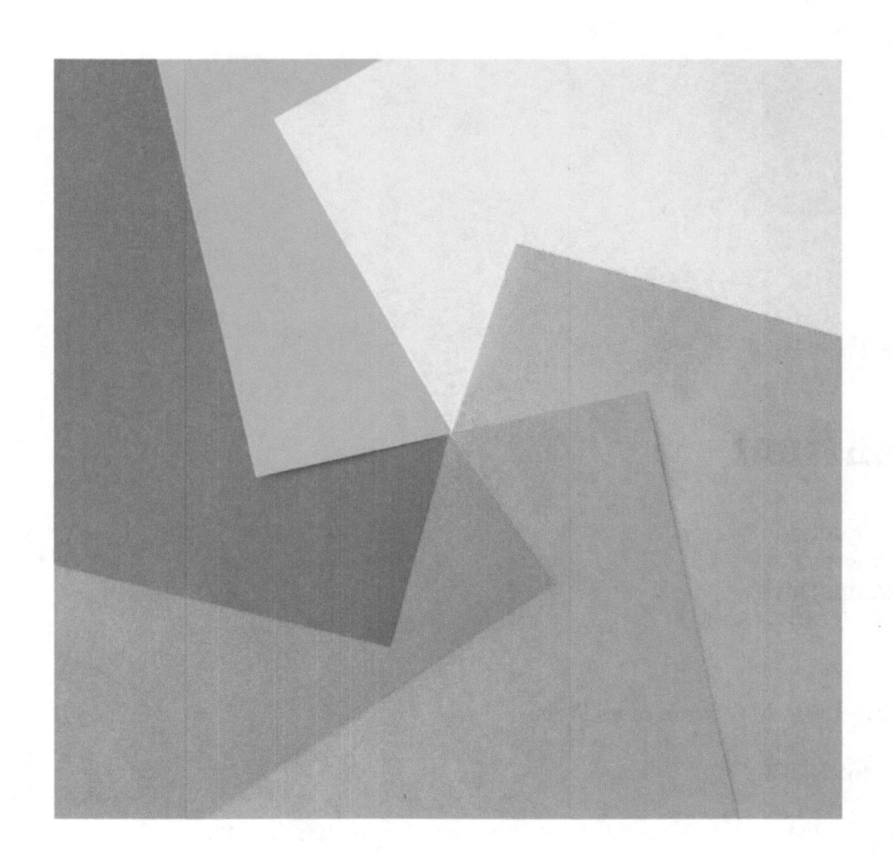

www.kendallhunt.com
Send all inquiries to:
4050 Westmark Drive
Dubuque, IA 52004-1840

ISBN 978-1-4652-7826-5

Printed in the United States of America

Dedication

This book is dedicated to my son, Max, who was born while I was writing this book, and to my husband, Andre, my biggest supporter in publishing this textbook.

Contents

Preface

This book is written with the students in my Informal Geometry course in mind. I have included extra information where I think it will be helpful based on my experience in working with students and their common misconceptions. Informal Geometry is a course designed for Education Majors. It is, however, the only one of the three courses in our Education sequence that is not restricted to Education Majors. Many of the students who take this course are Communications Media students and a few other students take the course to fulfill a Liberal Arts and Sciences requirement.

After giving certain definitions, I give instructions on how to read and understand them. Some definitions are by necessity dense, and it can be difficult to decipher them. Therefore, if you have trouble with a concept, read the rest of the paragraph to see if there is further explanation. I have also intentionally left vocabulary words in plain text, because research shows that if the vocabulary word is in bold, readers will only read the sentence the word is in. In this way, I hope to encourage the reader to read about the vocabulary word in context, thereby gaining a greater understanding of its definition.

The text contains both examples and non-examples where appropriate. It also includes an "Understanding the Text" section at the end of many sections of the text. Solutions to these problems are not included in this edition of the text. However, if you work through one of these problems and want to know whether or not your answer is correct, I am happy to look at your answer and let you know.

When writing this textbook, I focused on including many details that traditional textbooks typically gloss over. That said, there are likely places where I have not included as much explanation as I should have. Please let me know if, after reading a passage and wrestling with it for a while, you still do not understand the concept. It is possible that some element is missing from the passage that is important to your understanding. Alternatively, I may need to include more examples to clarify what is meant by a specific concept.

I care very much about my students' learning, and I hope that comes across to the reader. Please give me any comments and suggestions you have, and ask me any questions that arise as you read this book. Thank you.

Acknowledgment

Thank you to Fitchburg State University for the sabbatical that made this textbook possible, and many thanks to the students who read it, pointed out errors, and made suggestions.

About the Author

Amy Wangsness Wehe is an Assistant Professor at Fitchburg State University. Originally an Iowan, she earned her BA in Mathematics from Drake University in Des Moines, IA and PhD from Iowa State University in Ames, IA. She has been living in Leominster, MA since 2005 with her husband and now 2-year-old son.

Chapter 1

Points, Lines, and Planes

Points, lines, and planes are the basis for Euclidean Geometry, which is what this book is all about. Although there are other kinds of geometry, such as geometry on a sphere (imagine what a triangle would look like drawn on a basketball), we will concentrate on geometry you can draw on a flat piece of paper. The three-dimensional figures we consider will be extensions of that type of geometry.

1.1 Points

A point is zero-dimensional. This means it has no width or length. We represent a point with a dot or a small circle, although technically it is so small it cannot be seen.

• < −− This is a point.

Points are labeled with capital letters. Such as:

1.2 Lines, Rays, and Line Segments

Lines are one-dimensional objects. That means it has a length, but no width. The length is its one dimension. We represent a line with a picture like the following:

When we refer to this line, we call it Line $\overleftrightarrow{AB}$ because the points A and B are both on this line. Also, a line goes on forever in both directions, so it has arrows on both sides of it;

the symbol above the two points does, too. If there are many points labeled on a line, then there are many ways to name the line.

For instance, this line:

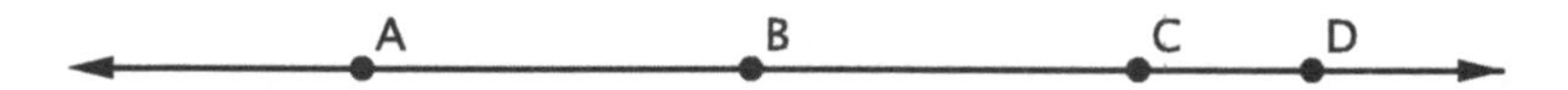

can be called Line $\overleftrightarrow{AB}$, Line $\overleftrightarrow{BC}$, Line $\overleftrightarrow{DA}$ and so on.
Sometimes lines are also labeled with lowercase letters, as in the picture below:

This line is Line m.

Notice that there are arrows on the ends of a line. This means that the line goes on forever in both directions, and this is what distinguishes a line from a line segment or a ray. A line segment such as this one:

has points on its ends, which is how the line segment is named. This line segment is $\overline{AB}$.
A ray only goes on forever in one direction:

This ray is called $\overrightarrow{AB}$. If the ray started at the point B instead, as in the picture below,

then it would be called $\overrightarrow{BA}$. As you can see from these examples, the order of the points in the name of a ray is important.

1.3 Planes

Planes are two-dimensional objects. They have both width and length. Examples of planes are the walls, ceilings, and floor of a room, except that planes go on forever in each direction.

A plane is typically pictured as a figure like the following:

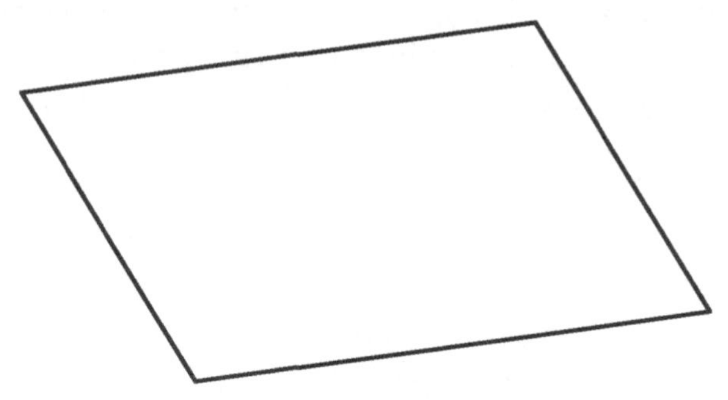

There is no requirement for there to be arrows on a plane, as there are in a line, indicating which directions it goes. However, a plane will extend in any of its directions forever, as indicated in the figure below:

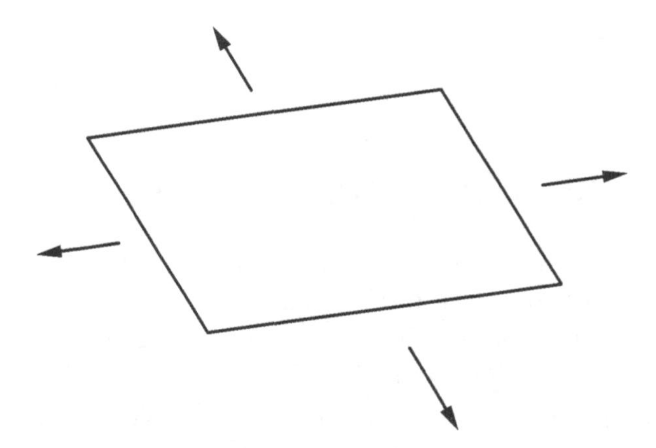

Planes are named by three points on the plane that cannot have a line drawn through them. Three points that can have a line drawn through all three of them are called collinear points. Therefore, it is said that a plane is defined by three noncollinear points contained by the plane.

For example, a plane like this one

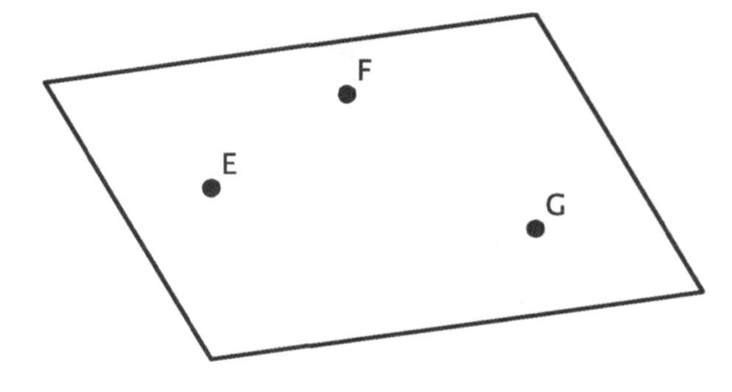

is named Plane EFG. You can see that the three points on this plane are noncollinear (that is, they do not lie on the same line) because a line that goes through any two of these points does not include the third point. Since any two points define a line, this is enough to show that these points are noncollinear. The following picture shows the lines that go through any two of the three points.

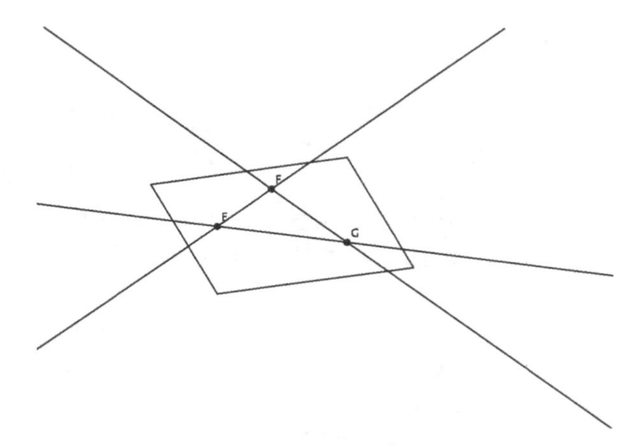

1.4 Collinear and Coplanar Points

Sometimes points are labeled in a picture, and sometimes they are not. A line has an infinite number of points on it, even though all of the points are not labeled in the pictures. In fact, you couldn't label all of the points on a line (or on a line segment, either), even if you tried. Likewise, given any two points, there is a line that goes through those two points, even if the line is not shown in the picture. For instance, in the picture below, the lines $\overleftrightarrow{EF}$, $\overleftrightarrow{AB}$, and $\overleftrightarrow{DC}$ are shown in the picture. The lines $\overleftrightarrow{EA}$, $\overleftrightarrow{FB}$, and $\overleftrightarrow{CB}$, however, still exist, even though they are not shown in the picture.

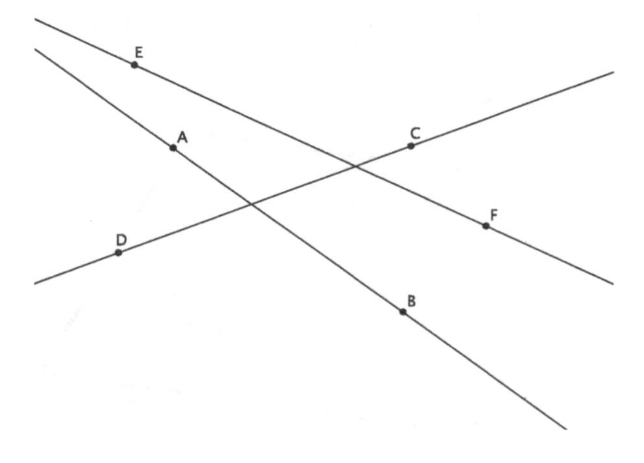

This becomes important when we begin talking about sets of points that are collinear or coplanar. For example, consider the picture below.

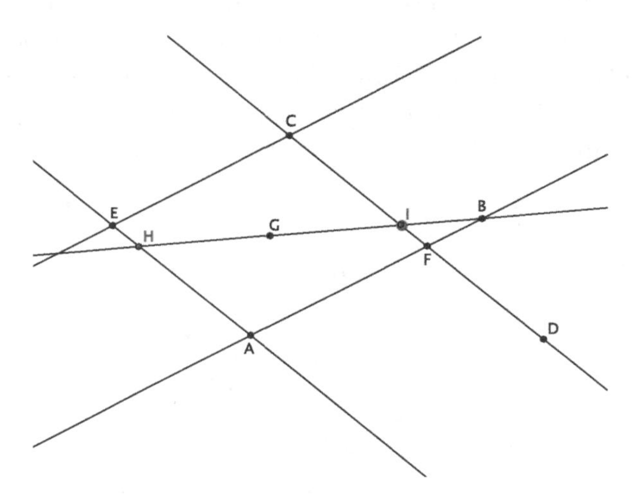

The points E, H, and A are collinear because they are all on the same line.

Again looking at the picture above, we see that the points C, I, and B are noncollinear. We can see this because the one line that goes through points C and I does not contain the point B. This is the easiest way to choose noncollinear points.

If we choose three points with no line pictured going through any two of the three points, we have to be careful. The points C, E, and B are definitely noncollinear, as you can see with your eyes that any line that goes through two of those three points cannot go through the third point. However, the points E, G, and F are collinear because there exists a line through all three of these three points, even though that line is not pictured. So that you can see this, I have included the line through the points E, F, and G in the picture below.

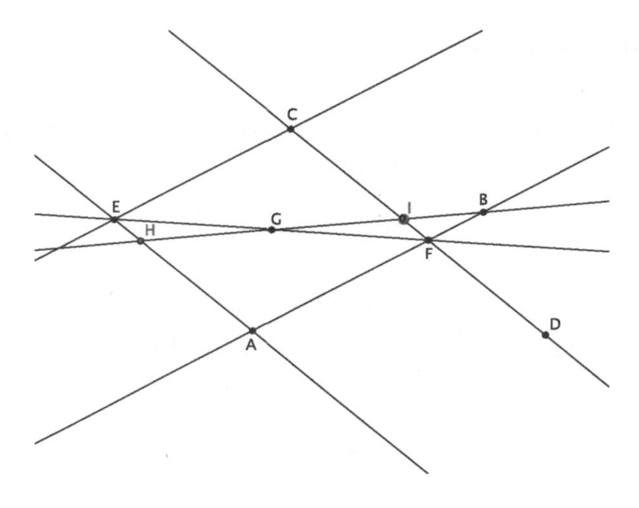

Notice a few things about a set of collinear points. First, the points are written as a list. We are not naming a line here; we are listing a set of points that are on that line. In addition, notice that there are three points in this list. Since any two points define a line, any two points are technically "collinear." That means the term is only interesting if we restrict the definition to three or more points. We can write the definition of collinear points in the following way: A set of three points or more points is collinear if there exists a line that contains all three points.

Understanding the Text:

Use the picture below to answer the following questions:

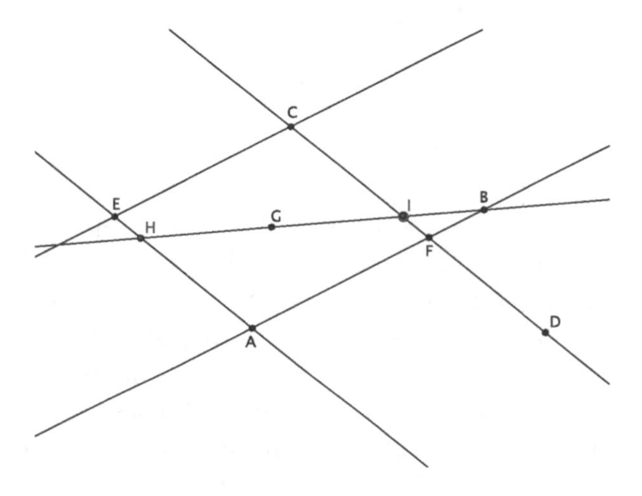

1. Name a set of collinear points not mentioned in the text above.

2. Name a set of noncollinear points not mentioned in the text above.

3. Do you see another set of three points that are collinear, but the line through them is not shown in the picture? If so, list them.

Next, we move on to coplanar and noncoplanar points. This is similar to the discussion of collinear points, but everything is increased by one dimension. As we already learned, three points define a plane, which is why we name planes using three points, such as Plane ABC. Therefore, we define a set of four points to be coplanar if there exists a plane that contains all four points. In the picture below, the points D, C, B, and A are coplanar.

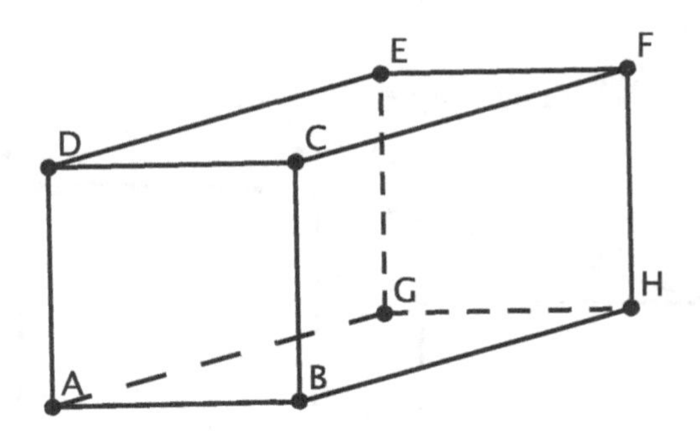

Likewise, we need a set of four points in order to have a set of noncoplanar points. The points A, B, C, and F in the figure above are noncoplanar. We can see the plane defined by the points A, B, and C, and the point F is clearly not on that plane, so these points are definitely noncoplanar.

Again, we have to be careful, because even if a plane is not shown in the picture, it still exists. Remember that the definition of coplanar is that there exists a plane that contains all four points. For instance, the points E, F, A, and B in the picture above are coplanar, even though the plane that goes through those four points is not pictured.

I have added that plane in the picture below:

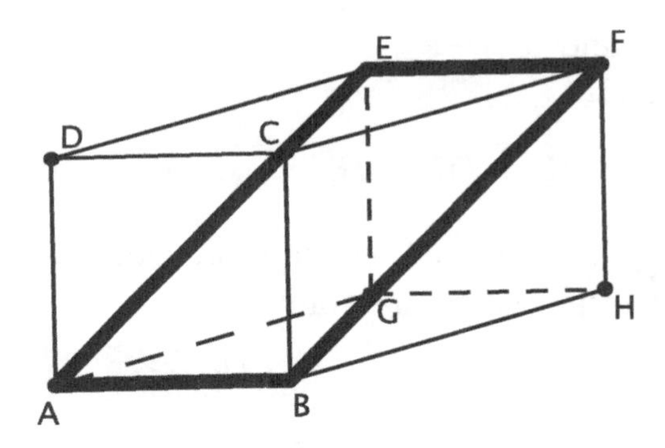

Another way to see this plane is to look at the walls of a room. Pick a wall in the room. Now, imagine a (very!) large piece of plywood in the room that reaches from the corners where your wall meets the ceiling to the corners where the wall opposite your wall meets the floor. This piece of plywood would span the room diagonally, and represents a plane equivalent to the one in the previous picture.

Understanding the Text:
Use the picture below to answer the following questions:

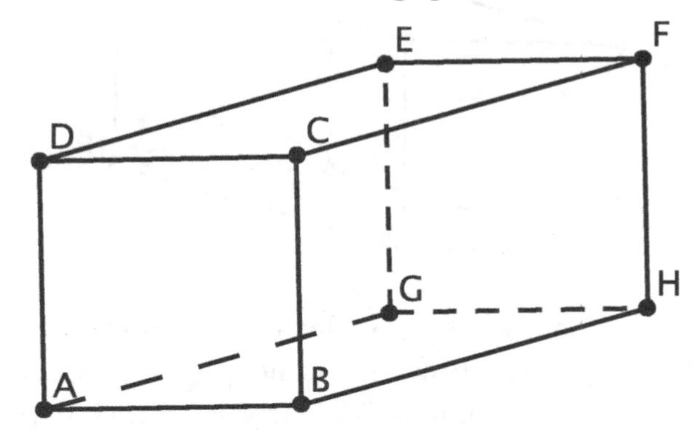

1. Name a set of coplanar points not mentioned in the text above.
2. Name a set of noncoplanar points not mentioned in the text above.
3. Do you see another set of four points that are coplanar, but the plane through them is not shown in the picture? If so, list them.

We will now expand our knowledge of lines and planes by looking at intersections of lines and planes and parallel lines and planes.

1.5 Intersections of Parallel Lines and Planes

Two lines intersect when they have at least one point in common. For example, in the picture below, the lines $\overleftrightarrow{HI}$ and $\overleftrightarrow{CI}$ intersect at the point I.

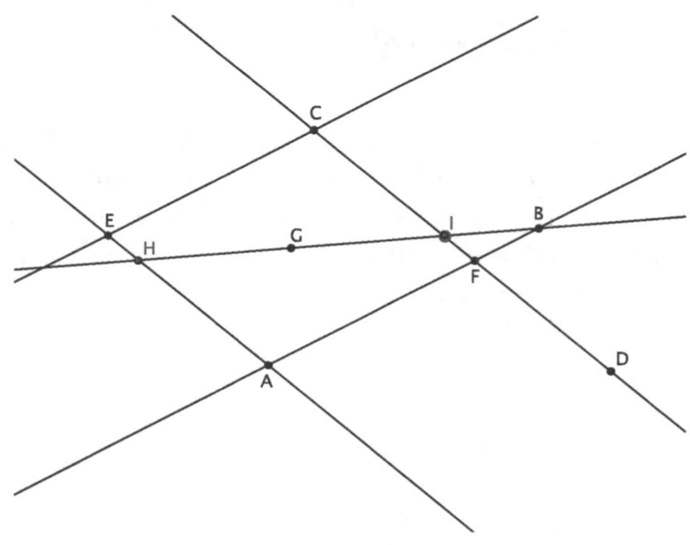

If two lines never intersect, they are called parallel. The lines $\overleftrightarrow{EA}$ and $\overleftrightarrow{CD}$ in the picture above are parallel lines because even if they are extended, they will never intersect.

Remember, just because a point is not shown in a figure does not mean that the point does not exist. If two lines will intersect when they are extended, they are not parallel lines. Remember, lines go on forever in both directions. Two lines are not parallel if they will *ever* intersect. The next picture illustrates this:

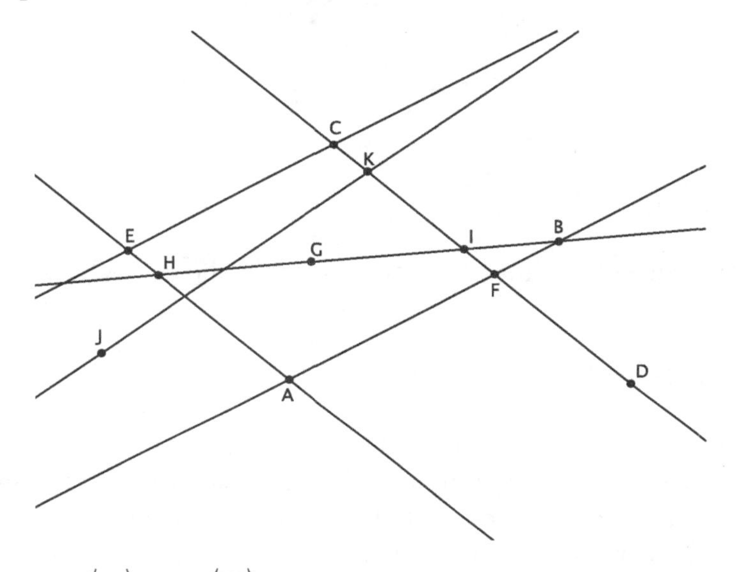

Notice that the lines $\overleftrightarrow{EC}$ and $\overleftrightarrow{JK}$ will intersect each other if they are extended. Therefore, these lines are NOT parallel.

Three lines are concurrent if all three lines intersect at one point. For example, in the figure below, the lines $\overleftrightarrow{AB}$, $\overleftrightarrow{FG}$, and $\overleftrightarrow{CD}$ all intersect at the point G. This makes these three lines concurrent.

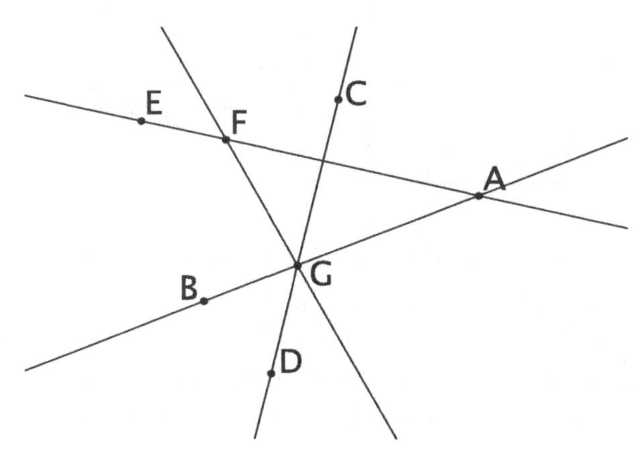

Understanding the Text:
Use the picture below to answer the following questions:

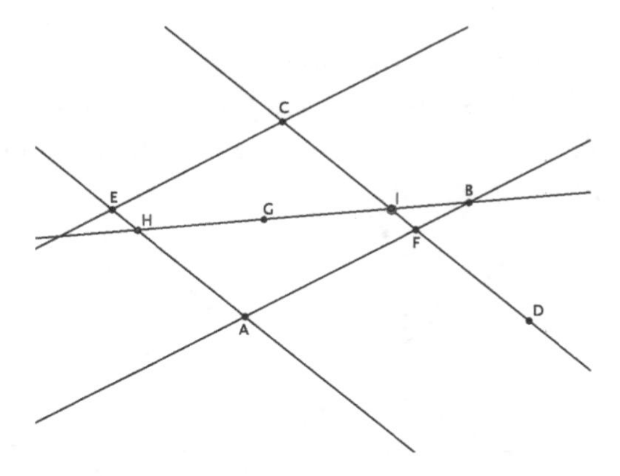

1. Name a pair of intersecting lines not mentioned in the text above, and name the point where they intersect.

2. Name a pair of parallel lines not mentioned in the text above.

We now look at how planes intersect each other. Again, two planes intersect each other when they have at least one point in common. Consider the planes, Plane DCBA and Plane DCFE, in the figure below. You can see that these two planes, as they are drawn, intersect at the line segment $\overline{DC}$. Now recall that planes go on forever in all directions. This means that the place where they intersect does not end at the points D and C, but instead continues on forever in all directions. In fact, Plane DCBA and Plane DCFE intersect at the line $\overleftrightarrow{DC}$.

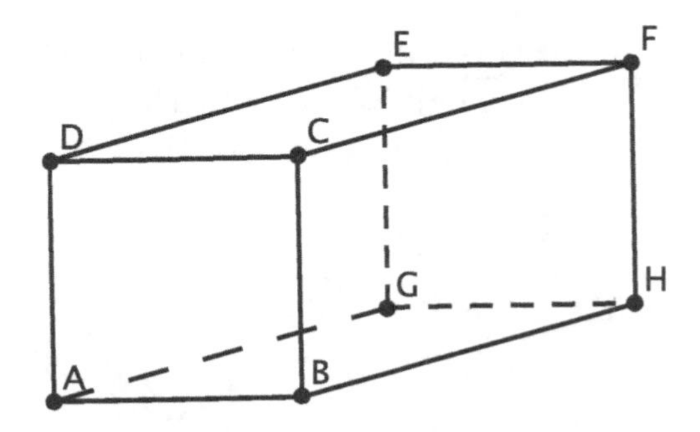

Remember that a line does not need to be drawn as a line in the picture in order to exist. This is another example of that. Although there is no "Line $\overleftrightarrow{DC}$" shown in the picture above, the line does exist and is the intersection of the two planes.

For a real life, 3-D example of the intersection of two planes, sit in a room and look at two adjacent walls in the room. Focus on where the two walls meet. The crease in the room where the two walls meet is their intersection. Now, imagine that the walls go on forever in all directions. Then the crease will go on in both directions forever. Therefore, the two planes meet at a line.

As with lines, two planes are parallel if they do not have any points in common. Again, just because two planes do not intersect in the picture does not automatically make them parallel. They are only parallel if they will **never** intersect, even if the planes that are pictured are extended forever.

Looking at the same picture of planes that we looked at before, shown again below, the planes Plane $DCBA$ and Plane $EFGH$ are parallel.

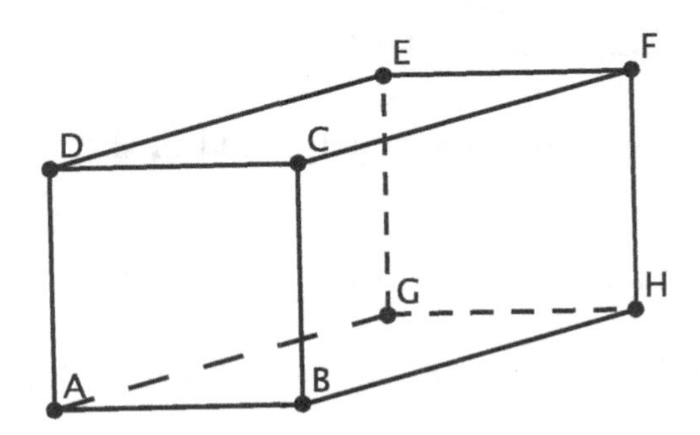

Looking again at a room, the plane containing the wall in front of you and the plane containing the wall behind you are likely to be parallel to each other. Likewise, the ceiling and the floor are likely to be parallel to each other. If you are in a room where the ceiling of the room is the roof of the house you are in, this may not be the case, since those ceilings are often slanted downward. Similarly, if you are in a very old house, it is possible that the house has shifted and the walls are not quite parallel to each other. If this is the case in the room you are sitting in, you may have to imagine a different type of room. Most classrooms are shaped like a box, and will work for this activity.

Two lines are skew lines if they are not in the same plane and they do not intersect. Recall that a pair of parallel lines must lie in the same plane. So, although skew lines do not intersect, they are not parallel, because they are not in the same plane. An example of skew lines in a room is the line between the floor and the front wall of the room, and the line between two walls at the back of the room. Those two lines are not on the same plane, but they also do not intersect. They are, therefore, skew lines.

Notice that two lines intersect at a point (if they intersect at all) and that two planes intersect at a line (if they intersect at all). When you go from lines to planes, you go "up a dimension"; you also go "up a dimension" when you go from the intersections of lines (points) to the intersection of planes (lines).

Understanding the Text:
Use the picture below to answer the following questions:

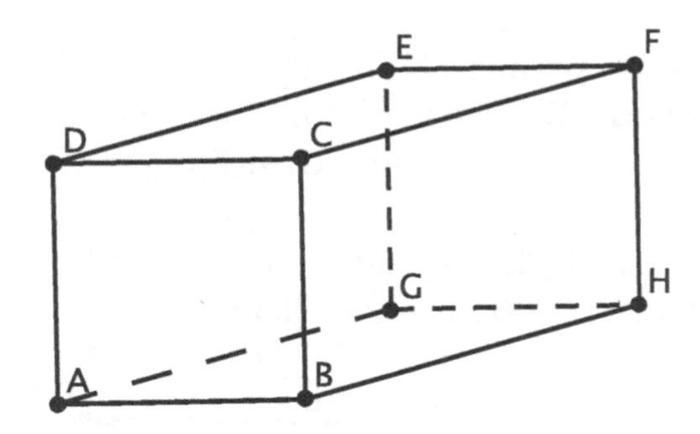

1. Name a pair of intersecting planes not mentioned in the text above, and name the line where they intersect.
2. Name a pair of parallel planes not mentioned in the text above.

In the next chapter, we will talk about angles and how angles relate to each other in particular situations.

Chapter 2

Rays and Angles

2.1 Definitions, Notation, and Measurement

Recall that a ray starts from a point and then extends forever in one direction, as in the picture below. A ray is labeled with two points on the ray, the starting point and another point on the ray, with an arrow over the top of the two points.

The ray pictured above is Ray $\overrightarrow{AB}$.

If you join two rays at their beginning points, you form an angle, such as the angle below. The point where the two rays are joined is called the vertex of an angle.

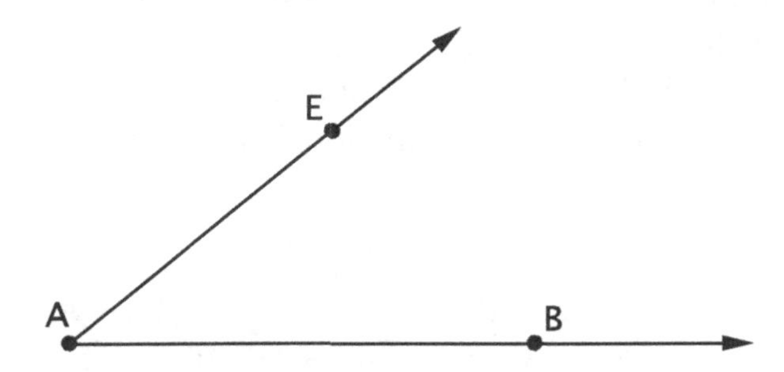

This angle can be called $\angle BAE$. The symbol in the front indicates you are naming an angle, and then the three letters are three points on the angle with the vertex of the angle listed in the middle and the other two points each on their own ray. Therefore, this angle can also be called $\angle EAB$.

An angle can be measured by a protractor. A short, simple YouTube video that shows you how to measure an angle with a protractor, called "Measuring Angles with a Protractor" can be found at `http://www.youtube.com/watch?v=50-9wgGufvc`[31].

Otherwise, I will explain how to use a protractor here. First, by example: we will measure $\angle BAC$ pictured below.

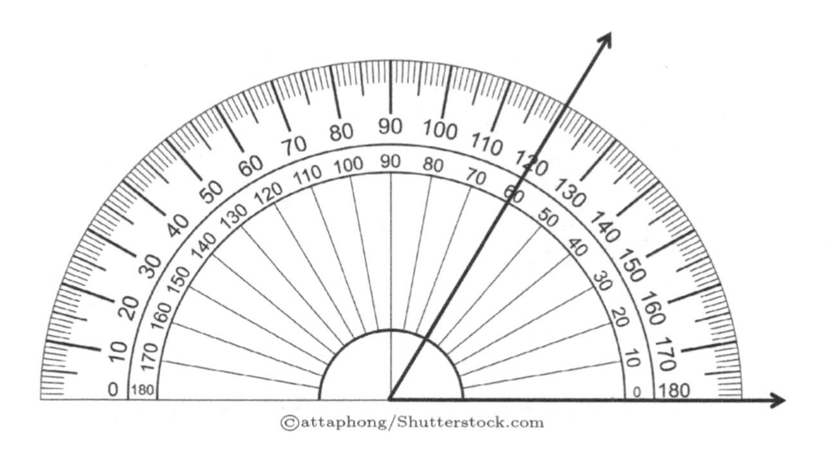

©attaphong/Shutterstock.com

The angle is already properly lined up on the protractor. The ray $\overrightarrow{AC}$ is lined up with the "0" line at the bottom of the protractor, and the vertex, A, is on the point at the center of that line. The ray $\overrightarrow{AC}$ is pointing toward the right, and crosses the numbers on the outside edge of the proctractor where the 0° mark is on the inside. Therefore, we will use the inside numbers of the protractor.

Now, we look at the ray $\overrightarrow{AB}$. This ray crosses the numbers along the edge of the protractor at the numbers 60 and 120. Since we are using the numbers on the inside, we will use the number 60. Therefore, the measure of $\angle BAC$ ($m\angle BAC$) is 60°.

Another way to think about which numbers to use is to think about whether or not the angle is larger or smaller than a right angle. Since this angle is smaller than a right angle, you must use the "60" instead of the "120." A right angle is a 90° angle.

For a second example, we will look at an angle where the ray points to the left.

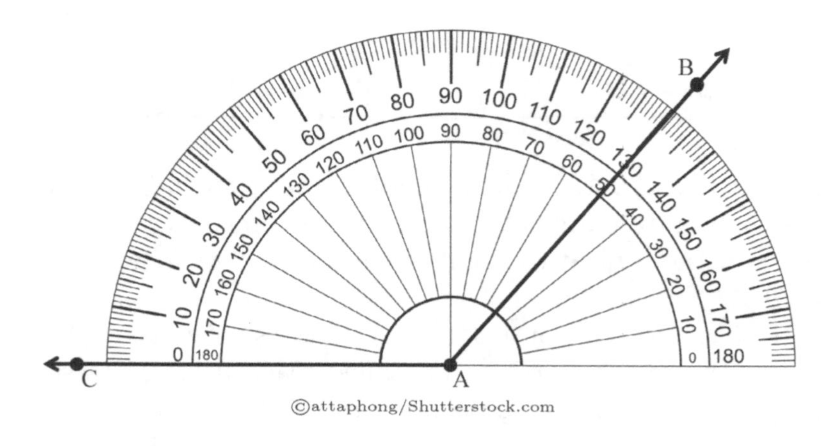

©attaphong/Shutterstock.com

Consider the angle BAC in the picture above. The ray $\overrightarrow{AB}$ is on the zero line at the bottom of the protractor, and Point A is on the point in the middle of the base line of the protractor, directly under the 90° mark. We see that the 0° the ray goes through is the outside number on the protractor. Therefore, we will read the outside numbers on the protractor. The ray $\overrightarrow{AC}$ goes through the numbers 50 and 130. Since we are using the outside numbers, we see that the measure of the angle is 130°.

Another way to think about which numbers to use is to think about whether or not the angle is larger or smaller than a right angle. Since this angle is larger than a right angle, you must use the "130" instead of the "50."

Different protractors measure angles slightly differently, but the main idea is that you line up one ray along the bottom edge (or along a specified line near the bottom edge) of the protractor with the vertex of the angle on the center of that edge (or line). Then you extend the other ray (either by drawing it longer or lining it up with a ruler) until it crosses the numbers written along the outside edge of the protractor. The angles in the examples above were already long enough to cross the numbers on the protractor, but not all of the angles you draw will be.

Now look at where the ray along the bottom of the protractor meets the numbers on the semicircle edge of your protractor. The numbers it meets should be 0° and 180°. Look at where the 0° meets your ray. Is it the inside or the outside number? If it is on the inside, use the inside numbers to measure your angle. If it is on the outside, use the outside numbers.

Now look again, where the other ray crosses the numbers on the outside of your protractor. Using the appropriate numbers (the inside numbers or the outside numbers), read off the number from your protractor. That is the measure of your angle.

If your angle measures 90°, it is called a right angle. If your angle measures less than 90°, it is called an acute angle. Above, $\angle BAC$ is an acute angle. If your angle measures more than 90°, it is called an obtuse angle. The second example, $\angle DEF$ measures an obtuse angle.

Understanding the Text:

1. State the degree measure of $\angle DEF$ on the protractor below.

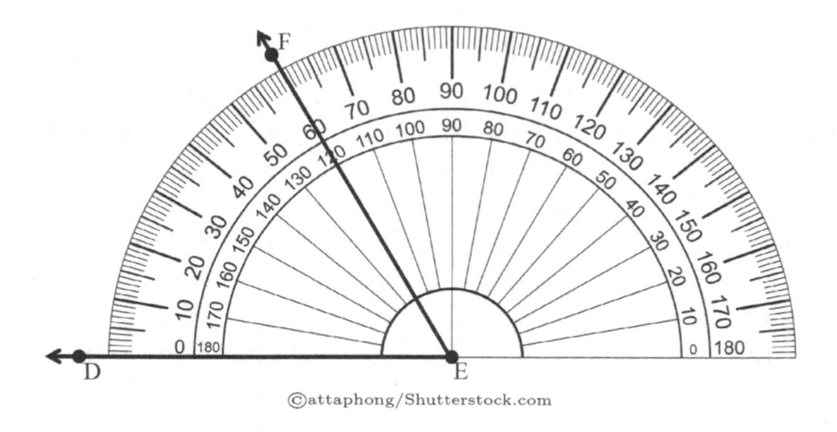

2. Is $\angle DEF$ in Problem 1 a(n) acute, obtuse, or right angle?

3. Measure the angle below using your own protractor.

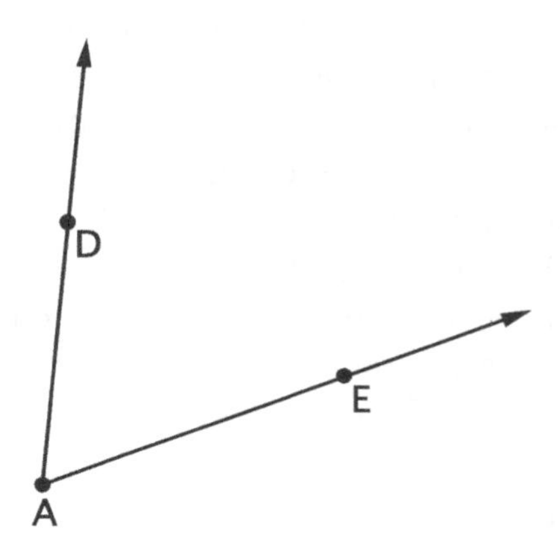

2.2 Special Pairs of Angles

We will now look at some special angles and some special relationships between angles.

We start with two parallel lines. You can make two parallel lines using only a straight edge (a ruler without using the measurement markings) and a compass; for instructions on how to do this, see Appendix A.1.

Once you have two parallel lines, make a transversal through the two lines. A transversal is a line intersecting two parallel lines. In the picture below, lines l and m are parallel lines, and line p is a transversal.

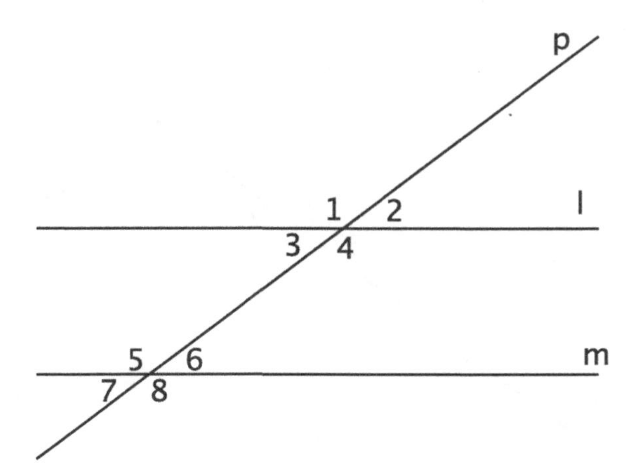

Certain pairs of angles produced by two parallel lines and a transversal have special names. Vertical angles are angles that are across from each other where two lines intersect. In the picture above, $\angle 1$ and $\angle 4$ are one example of a pair of vertical angles.

Alternate exterior angles are angles that are on opposite sides of the transversal and are outside of the parallel lines. In the picture above, $\angle 1$ and $\angle 8$ are one example of a pair of alternate exterior angles. Exterior angles on the same side of the transversal are just what they say they are. They are on the outside of the parallel lines, but are on the same side of the transversal. In the picture above, $\angle 1$ and $\angle 7$ are one example of a pair of exterior angles on the same side of the transversal.

Alternate interior angles are angles that are between the parallel lines and are on opposite sides of the transversal. In the picture above, $\angle 3$ and $\angle 6$ are one example of a pair of alternate interior angles. Interior angles on the same side of the transversal are angles that are on the same side of the transversal and between the parallel lines. In the picture above, $\angle 3$ and $\angle 5$ are one example of a pair of interior angles on the same side of the transversal.

If you think of the angles as sets of four, one set touching one parallel line and the other set touching the other parallel line, then corresponding angles are in corresponding positions in their respective set of angles. In the picture above, $\angle 1$ and $\angle 5$ are one example of corresponding angles.

Two angles are supplementary angles if together they form a straight line. In the figure above, $\angle 1$ and $\angle 2$ are supplementary angles. It is not necessary for parallel lines to be present in order to have supplementary angles, however. In the figure below, $\angle 1$ and $\angle 2$ are supplementary angles because together they form the line l. Similarly, $\angle 1$ and $\angle 3$ are supplementary angles because together they form the line m.

Two angles can be supplementary even if they do not touch each other in the picture. Take two angles and line them up so that a ray of one angle coincides with a ray of the other angle. If other rays of the angles together form a straight line, then those two angles are supplementary.

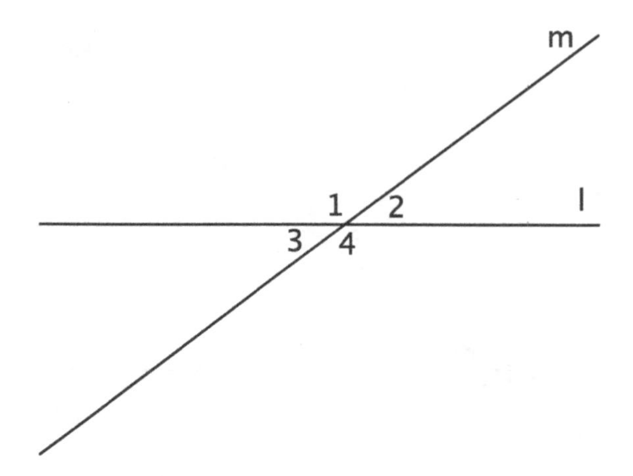

Try measuring the angle formed by $\angle 3$ and $\angle 4$ together on your protractor. What is the measure of the angle? You should notice that this angle is a straight line, since $\angle 3$ and $\angle 4$ together form the line l. If you line one ray of this angle up with the $0°$ line on your protractor, you should notice that the other ray of this angle also lies on this line. Starting with $0°$ on one side of the protractor, the other ray should be on the $180°$ line. Therefore, the angle measure of a straight line is $180°$.

Perhaps you know that if go all the way around a circle, you get an angle measuring $360°$. This is why if you spin your car all the way around so you are facing in the same direction you started, it is called "doing a 360." From the picture above, you see that $\angle 1$ and $\angle 2$ form a $180°$ angle, and $\angle 3$ and $\angle 4$ also form a $180°$ angle. If you put these two $180°$ angles together, you get the full circle around the intersection point of lines m and l. Also, if you add the measure of the two angles, you get $360°$, as you may have guessed.

Another important type of angles are complementary angles. Complementary angles are two angles whose measures sum to $90°$. In the picture below, $\angle 1$ and $\angle 2$ are a pair of complementary angles. Just as with supplementary angles, the two angles do not need to be touching each other in the picture in order to be complementary angles. If two angles would form a right angle when moved together so that the ray of one coincides with the ray of the other, the angles are complementary angles.

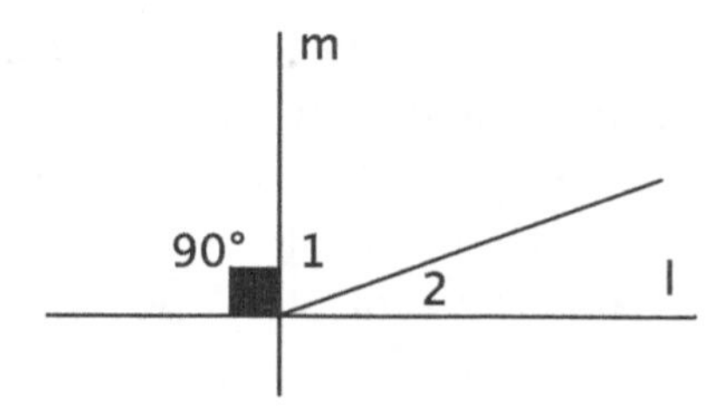

Notice that $90°$ is half of $180°$. Look at your protractor. What line is half-way around the semicircle your protractor makes? Hopefully, you have found the only line along the outside of your protractor that only has one number on it. That is because that line has the same measure whether you start on the right side or on the left side of your protractor when measuring an angle.

Draw a line segment on a piece of paper. Now draw another line segment through that line that forms a 90° angle with that line. To do this, first make a small point on your line segment and call it Point A. Then put your first line on the 0° line with point A on the center point of the 0° line on your protractor. Then, make a small mark at the 90° mark on your protractor and call that point B. Remove your protractor and use a straight edge to make a line segment containing (including) the points A and B. These two line segments form a 90° angle with each other. In fact, all four of the angles formed by your two line segments are right angles (90° angles). Consider the lines containing your two line segments (imagine that your line segments are extended forever in each direction). Two lines that form 90° angles with each other are called perpendicular lines.

There are a couple of other ways you can create two lines that are perpendicular to each other, depending on what tools you have at your disposal at a particular time. If you have thick paper with sharp corners, you can draw a perpendicular line by tracing the edge of the sharp corner of the piece of paper with one edge placed along the first line you drew. There are also instructions in Appendix A.1 on how to draw a line perpendicular to a given line using only a straight edge (a ruler without measurements) and a compass.

Understanding the Text:

1. On a piece of paper, draw a line and then draw a line intersecting that line. Choose two angles formed by these two lines that are adjacent (next to) each other; that is, two angles that have a ray in common. Label one of the angles "$\angle 1$" and one of the angles "$\angle 2$." Measure both $\angle 1$ and $\angle 2$ with your protractor and then add the two measures together. What kind of angles is this pair of angles? Was your answer what you expected it to be? If not, try measuring them again.

2. On a piece of paper, draw a line and then draw a line perpendicular to that line. A perpendicular line is a line that meets the original line at a 90° angle. Next, draw a ray beginning at the point where your two perpendicular lines intersect and call this ray $\overrightarrow{k}$. Ray $\overrightarrow{k}$ will form two angles with your perpendicular lines. Label these angles $\angle 3$ and $\angle 4$, as in the figure below.

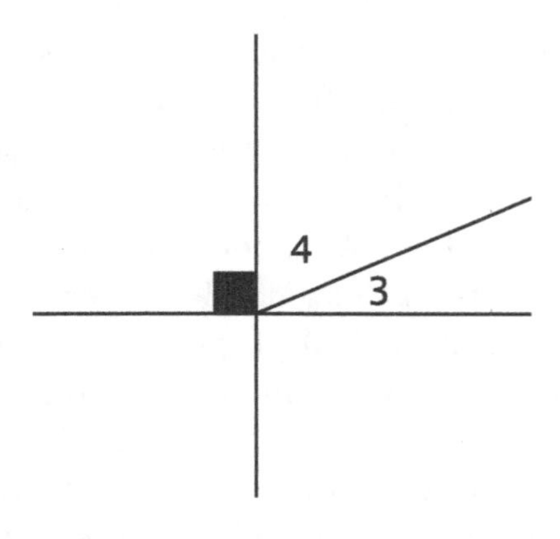

   (a) Measure both $\angle 3$ and $\angle 4$ with your protractor and then add the two measures together. Was your answer what you expected it to be? If not, try measuring the angles again.

   (b) The angles $\angle 3$ and $\angle 4$ are a pair of angles with a special name. What kind of angles are they?

In the next chapter, we will look at various polygons, as well as special angles created by those polygons.

Chapter 3

Triangles

3.1 Triangles and Their Special Angles

A triangle is a figure created by three noncollinear points (called vertices) and all possible line segments between those points. A triangle will have three sides (the line segments between the vertices) and three angles (created by its vertices and its sides). Its name, tri/angle, comes from the fact that it has three angles. See the figure below:

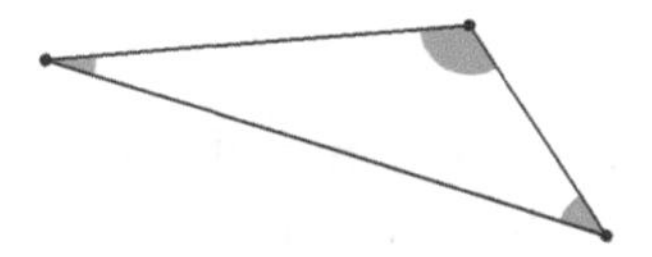

A triangle need not have its vertices explicitly showing. The vertices of a triangle are located at the intersections of its sides, whether they are shown or not, as in the picture below.

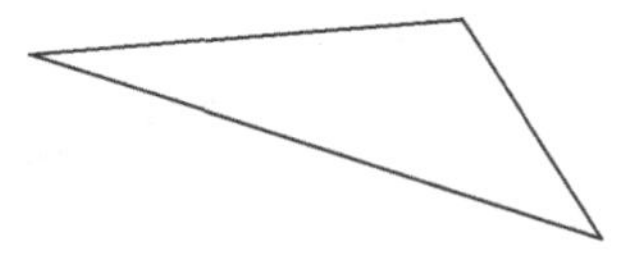

You may say, "Hey! There are no rays in a triangle!" If so, good for you! You are correct, and it was astute of you to remember that you need two rays to make an angle. However, remember when we talked about collinear points and coplanar points (Section 1.4)? The lines and planes exist, even though they are not shown in the picture. A similar thing is happening here. The vertex angles of a triangle are defined by using the vertex of a triangle as the vertex of an angle and rays containing the sides of the triangle coming out of that vertex as the rays of the angle.

A vertex angle of the triangle below is labeled α, pronounced "alpha."

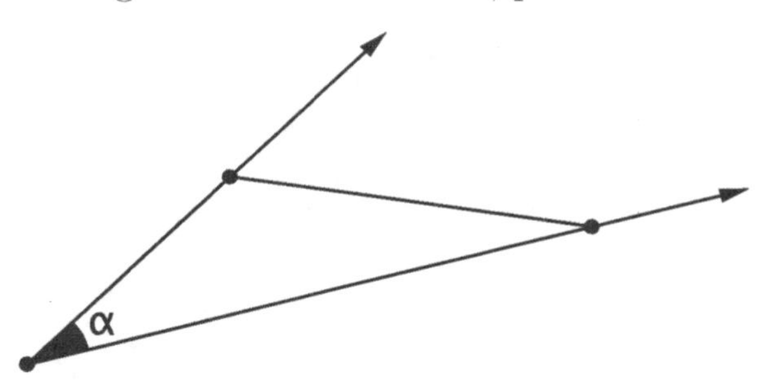

Clearly, a *vertex angle* is called a vertex angle because its vertex is at the *vertex* of the triangle.

There are several other special angles associated with a triangle. The exterior angle is an angle that is supplementary to the vertex angle of a triangle. We learned in Section 3.1 that two angles are called supplementary angles if together the angles create a straight line. In the picture below, the exterior angle, β, pronounced "beta," is supplemental to angle α.

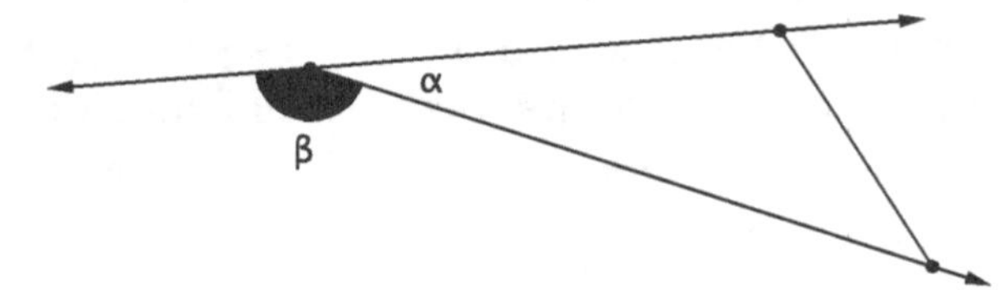

You may have noticed that $\angle\alpha$ in the picture above has another exterior angle associated with it. To find that angle, create the line containing the other ray of $\angle\alpha$ and find the angle supplementary to $\angle\alpha$ on that line. This is shown in the picture below. Again, $\angle\alpha$ is the vertex angle and $\angle\beta$ is the exterior angle associated with it.

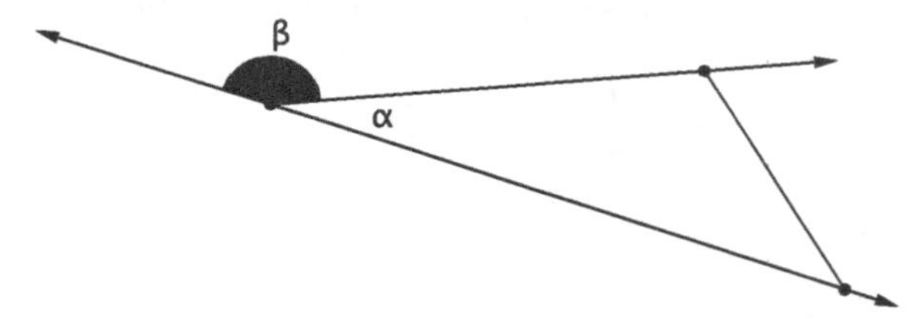

If a triangle is a regular triangle, then it also has an angle called a central angle. A regular triangle, also known as an equilateral triangle, is a triangle in which all of the sides are the same length and all of the angles are the same size. Since "lateral" is another word for "side" and "equi" is another way of saying "equal," the word equilateral means "equal sides." You might wonder why the word says nothing about equal angles. Well, it turns out that if all of the sides of a triangle are the same length, then all of the angles are the same size also. A word of caution: this is not the case for other shapes with equal sides.

The picture below is of an equilateral triangle.

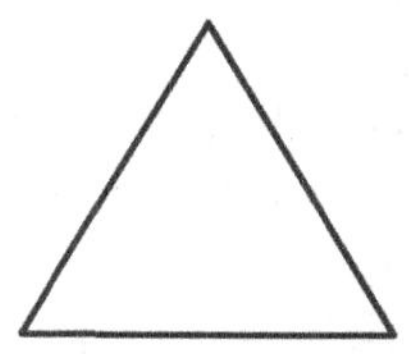

A central angle of an equilateral triangle is called a central angle because its vertex is at the center of the triangle. The rays of a central angle go through two (adjacent) vertices of the triangle. In the equilateral triangle below, $\angle\alpha$ is one example of a central angle.

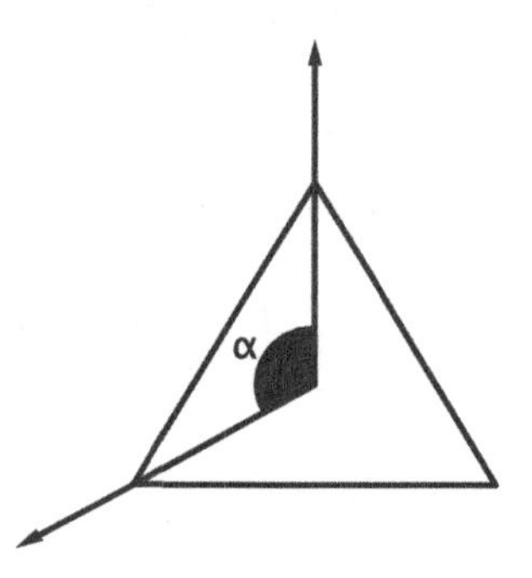

A triangle with all sides equal (and all angles equal) is called an equilateral triangle. A triangle with a vertex angle that is a right angle is called a right triangle. A triangle with a vertex angle that is obtuse (measures greater than 90°) is called an obtuse triangle. A triangle in which all vertex angles are acute (measures less than 90°) is called an acute triangle. A triangle in which no two sides are the same length is called a scalene triangle. In a scalene triangle, the angles also each have a different measure. An isosceles triangle has two sides the same length. An equilateral triangle is an isosceles triangle, but an isosceles triangle is not necessarily an equilateral triangle (the third side could have a different length).

In an isosceles triangle, the two angles opposite the sides with the same length are the same size (have the same measure).

Understanding the Text:

1. Draw an equilateral triangle. Next, sketch all of the central angles of the triangle. How many central angles are there?
2. Draw a triangle. Next, sketch all of the vertex angles of the triangle. How many are there?
3. Draw a triangle. Next, sketch all of the exterior angles of the triangle. How many are there?

4. Draw an example of an isosceles triangle, an equilateral triangle, a scalene triangle, and acute triangle, an obtuse triangle, and a right triangle.

5. Draw a triangle on a piece of scratch paper. Color or shade in the tips of the triangle. Cut off all three corners of your triangle so that the pieces are only partially shaded in. Take the pieces you cut off of your triangle and lay them on a hard, flat surface (such as a table) edge to edge, with the shaded corners touching each other. If you combine all three angles, what angle do they make? If you added the measures of all of the angles of your triangle together, what would you get?

3.2 Special Angle Measure of Equilateral Triangles

We can find the angle measure of a central angle of an equilateral triangle in the following way. First, draw all of the central angles of the equilateral triangles, as in the picture below.

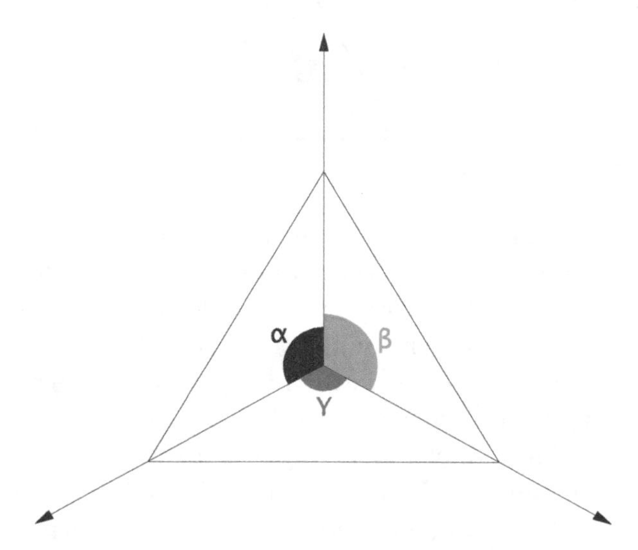

Notice that there are three central angles in the triangle. Because an equilateral triangle is a regular triangle, all of the central angles are the same size. The measure of an angle that goes all around the circle is 360°. Remember, "doing a 360" means rotating all the way around the circle. So, each of these three angles are the same size, and their measures add up to 360°. Another way to think about it is that these three angles divide the "full circle" angle into three equal parts. So, what is the measure of each of the angles? Take a moment to think about it and see if you can figure it out.

OK, here is the answer: if 360° is divided into three equal measures, that would be $360/3 = 120$. Therefore, the measure of a central angle of an equilateral triangle is 120°.

Next, let's see if we can find the measure of the vertex angles of an equilateral triangle. If you have not yet done so, do the Understanding the Text Question #5 in Section 3.1. Once you have done the activity in that question, you should have some evidence that the sum of the measures of the vertex angles of any triangle is 180°. Given that in any triangle, the sum of the measures of the vertex angles is 180°, we now want to find the measure of each angle in an equilateral triangle. Recall, again, that in an equilateral triangle, all of the angles are equal. Therefore, we have three angles of the same size that add up to 180°. Equivalently, a measure of 180° must be divided into three equal pieces. Therefore, each of the three angles in an equilateral triangle must have measure $180°/3 = 60°$, as shown in the picture below.

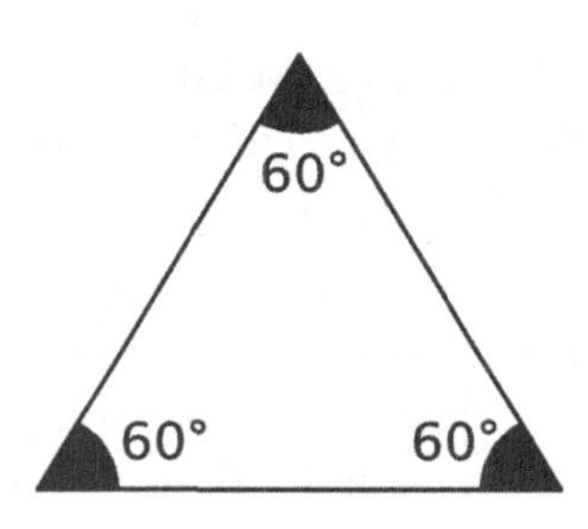

Remember, you can only say for sure the measures of the vertex angles of an *equilateral* triangle. For any other type of triangle, you would need to measure at least some of the vertex angles with a protractor in order to figure out the measure of all of them.

Finally, let's look at the measure of the exterior angles of an equilateral triangle. What do we know about exterior angles? Recall how they are related to the vertex angles of a triangle. Together, the exterior angles and the vertex angles make a straight line. That means that the sum of the angles is 180°. Written as a formula, that says:

$$\text{vertex angle} + \text{exterior angle} = 180°$$

Since we already know the measure of the vertex angles of equilateral triangles is 60°, this formula becomes: $60° + \text{exterior angle} = 180°$. Therefore, the measure of the exterior angle of an equilateral triangle is $180° - 60° = 120°$. This can be seen in the picture below.

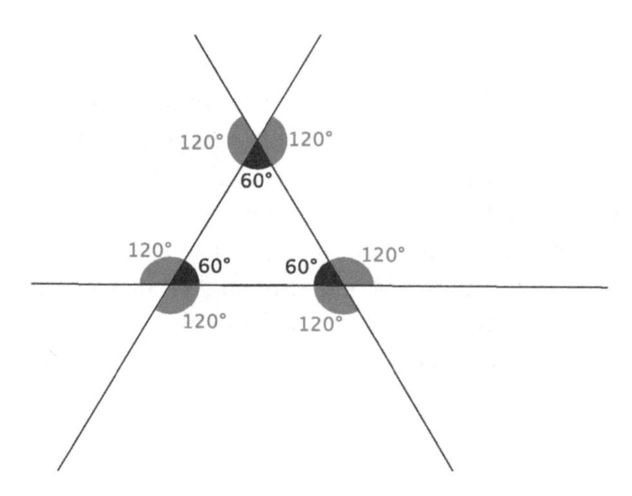

You may have noticed that the measure of the central angle of an equilateral triangle and the measure of the exterior angle of an equilateral triangle are the same. This is not a coincidence. Notice that the measure of a central angle of an equilateral triangle is 120° and that the vertex angle of an equilateral triangle is 60°. Therefore, the relationship between the measure of these angles in an equilateral triangle is: central angle + vertex angle = $120° + 60° = 180°$. Notice that this means those angles are supplementary to each other. You may also recall that a vertex angle and an exterior angle adjacent to it are supplementary angles. Take a minute to convince yourself that since an exterior angle of an equilateral triangle and a central angle of an equilateral triangle are both supplementary to a vertex angle of an equilateral triangle, they are equal to each other.

We will see more of this in the next chapter when we talk about quadrilaterals and other polygons.

Understanding the Text:

1. Given a scalene triangle, what is the least number of angles you would need to measure in order to determine the measure of all three angles of the triangle?

2. Given an isosceles triangle that is not also an equilateral triangle, what is the fewest number of angles you would need to measure in order to be able to determine the measure of all three angles of the triangle? You may assume that you can tell which angles are congruent, and you may also specify which angles of the triangle must be measured.

3.3 Median, Altitude, and Angle Bisectors

The midpoint of a line segment is the point on the line segment that is an equal distance from the endpoints of the triangle. In other words, given a line segment $\overline{AB}$, the point C

is the midpoint of $\overline{AB}$ if C is on line segment $\overline{AB}$ and the line segment $\overline{AC}$ has the same length as the line segment $\overline{CB}$. See the picture below.

The figure resulting from drawing all of the line segments from a vertex to the midpoint of the opposite side of an equilateral triangle looks like this:

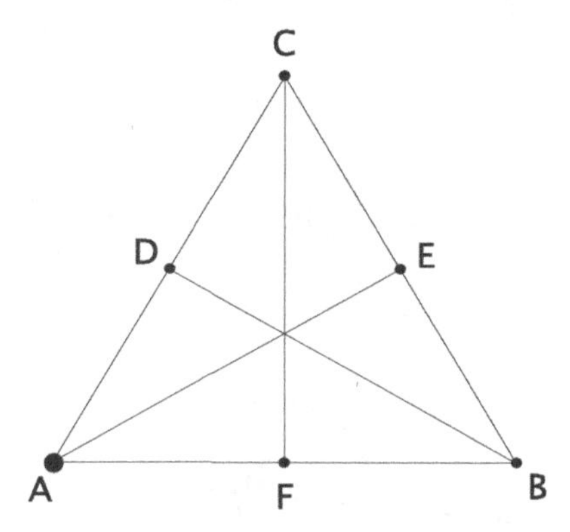

The points D, E, and F are the midpoints of the sides of the triangle. The center of the triangle is the point where all of these line segments meet. This is how you find the center of an equilateral triangle in order to draw a central angle.

Each of the three lines drawn in the equilateral triangle above is called a median of the triangle. A median can be drawn in any triangle. It need not be an equilateral triangle. In order to draw a median, take one side of a triangle and find its midpoint. Then draw a line segment from the midpoint of the edge of the triangle to the opposite vertex of the triangle (that is, to the vertex that is not an endpoint of that edge). This line segment is a median of the triangle.

The altitude of a triangle is the line segment from the vertex of a triangle to the line containing the opposite side, which makes a right angle with the line.

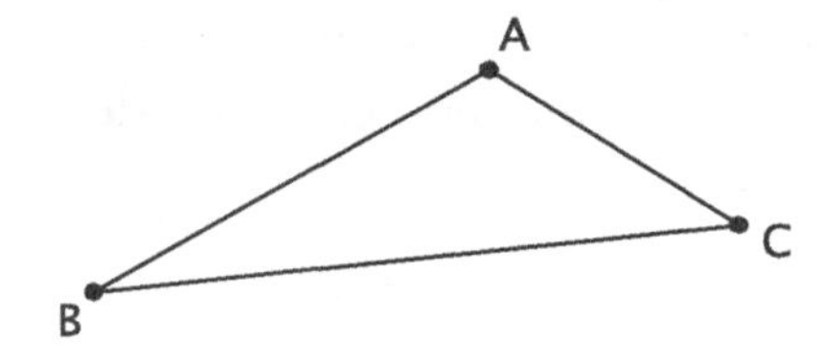

To find the altitude of the triangle above through point A, first draw line $\overleftrightarrow{BC}$.

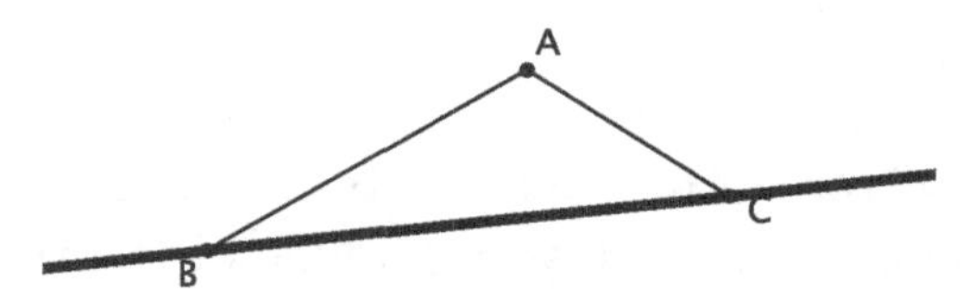

Then draw a line through point A perpendicular to line $\overleftrightarrow{BC}$.

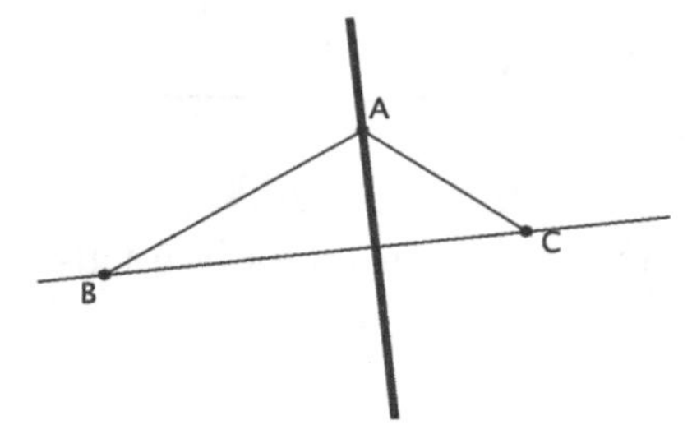

Call the point where the perpendicular line crosses $\overleftrightarrow{BC}$ Point D.

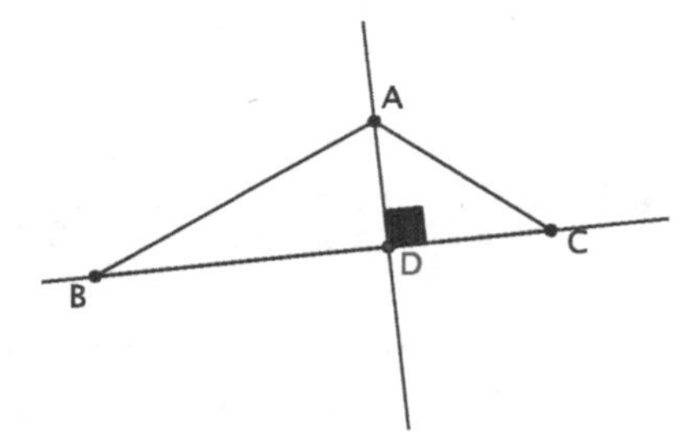

The altitude of the triangle through point A is $\overline{AD}$.

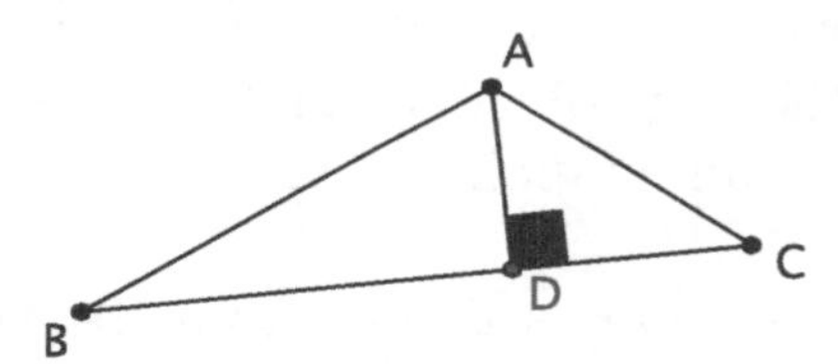

The angle bisector of a triangle is a line segment from the vertex of a triangle to the opposite side that bisects (divides in half) the vertex angle at that vertex.

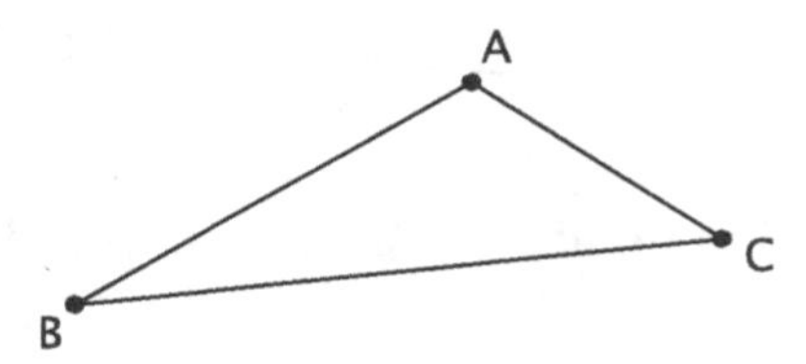

In order to find the angle bisector of the triangle above through the point A, begin by using a protractor to draw the line bisecting the vertex angle of the triangle at the point A.

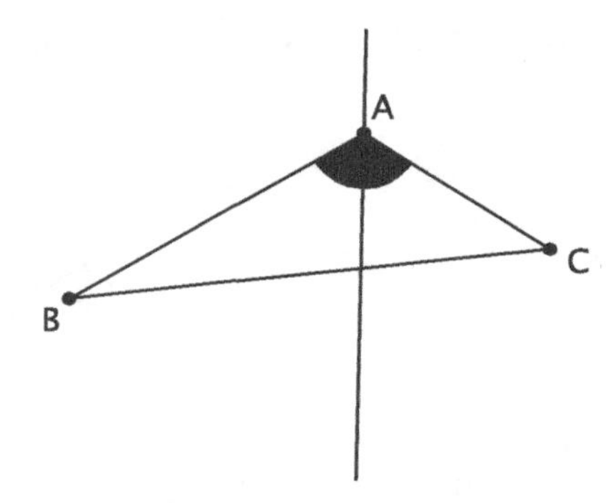

Call the intersection of that line and the line segment $\overline{BC}$ Point D.

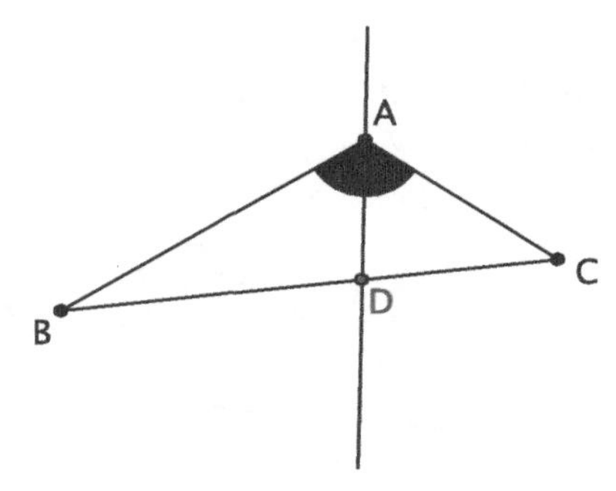

The angle bisector of $\triangle ABC$ at vertex A is AD.

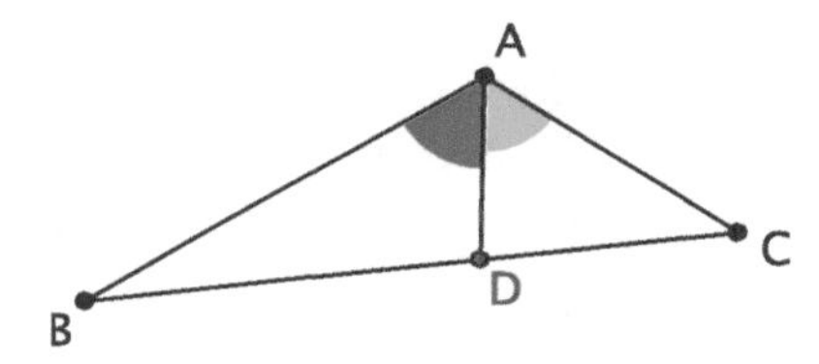

See Appendix A.4 to see how to bisect an angle using only a straight edge and compass.

Understanding the Text:

1. Explore on your own whether or not the medians of any given triangle meet at one point by drawing a variety of different triangles, measuring the midpoints of all the sides accurately, and then drawing all of the medians of the triangle. Do all of the medians of each triangles meet at one point?

2. Find the altitude of the triangle below through the point B.

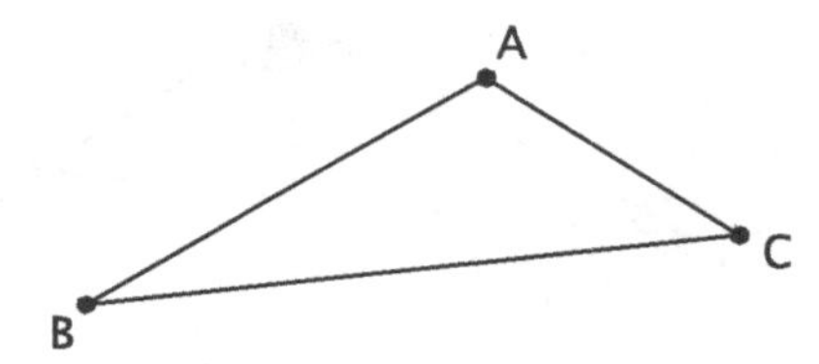

Notice that this altitude is not inside the triangle. This demonstrates why you need to draw the line through the opposite side, and cannot necessarily simply use the line segment itself.

3. Find all of the angle bisectors of the triangle below. Measure the angles accurately when you bisect the angles.

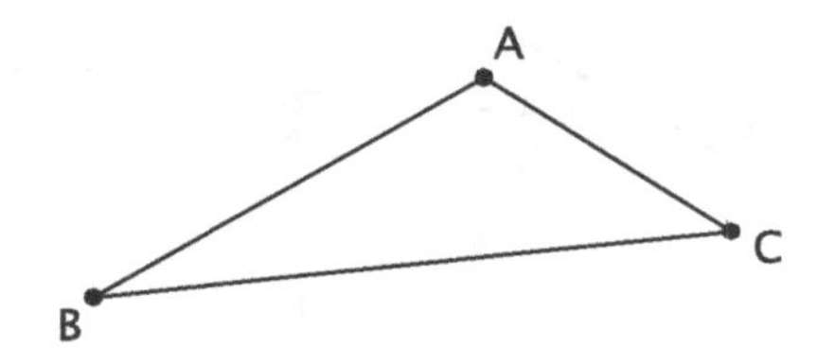

Chapter 4

Quadrilaterals and Special Angles

4.1 Quadrilaterals

A quadrilateral is a figure created by four points, called vertices, no three of which are collinear, where each vertex is connected to exactly two other vertices by a line segment, called a side.

The pictures below are examples of quadrilaterals:

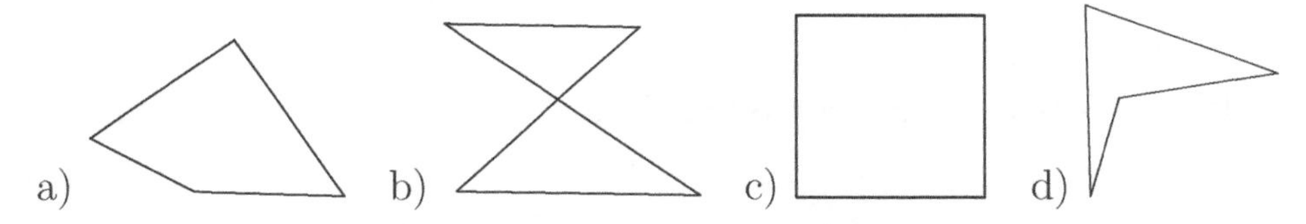

The following pictures are examples of four-sided figures that are not quadrilaterals:

Three vertices are collinear. Vertex 2 and 4 are connected to three vertices.

A simple quadrilateral is a quadrilateral with edges that intersect only at their endpoints. In a simple quadrilateral, the edges do not cross each other. See an example of a simple quadrilateral and an example of a quadrilateral that is not simple below.

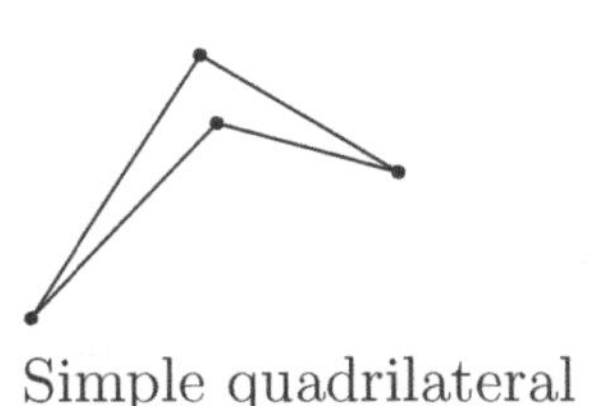

Simple quadrilateral

Not a simple quadrilateral: the edges intersect.

A simple quadrilateral has an "inside" and an "outside" of the quadrilateral. Since the sides do not cross, each other, they create a "fence" that contains a space called the "inside" of the quadrilateral. If a quadrilateral is not a simple quadrilateral, it is called a complex quadrilateral. In elementary school, we learn only about a "quadrilateral," which is what we will call a simple quadrilateral.

A convex quadrilateral is a simple quadrilateral for which any line segment between two points in or on the quadrilateral stays completely within that quadrilateral. The following quadrilateral is a convex quadrilateral.

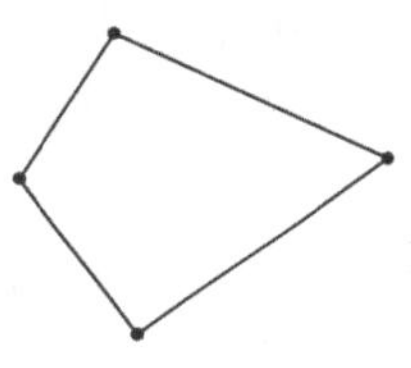

The following quadrilateral is not a convex quadrilateral:

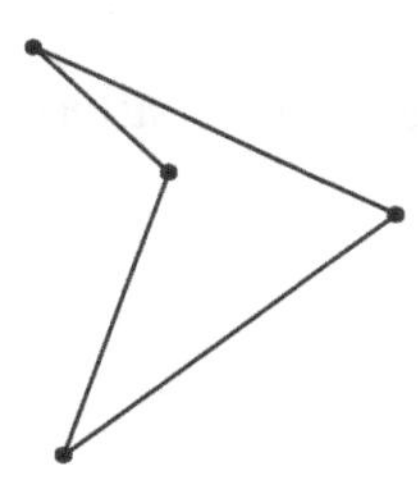

This is because there is at least one line segment between two points of the quadrilateral that does not stay completely within the quadrilateral. The line segment $\overline{AB}$ below has endpoints A and B, each of which are inside the quadrilateral, but the point C on the line segment is one example of a point on the line segment $\overline{AB}$ that lies outside of the quadrilateral.

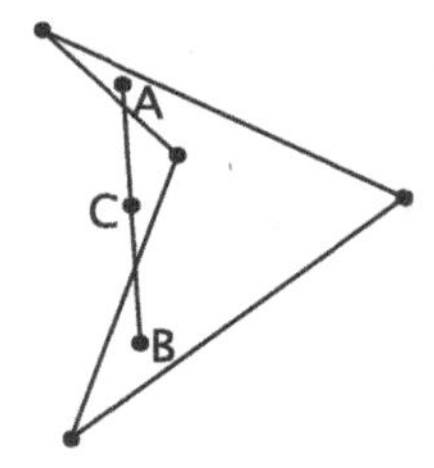

Just as there is a regular triangle (an equilateral triangle), there is a regular quadrilateral. Recall that a "regular triangle" was defined to be a triangle with all equal sides and all equal angles. The same is true of a regular quadrilateral: it is a convex quadrilateral with four sides the same length and four angles equal in size. A regular quadrilateral also has a special name. What is it?

Remember that any triangle with three sides the same length also has three angles the same size. This is not the case with quadrilaterals. A rhombus is a simple quadrilateral with all four sides equal in length. The angles of a rhombus are not necessarily all equal.

Regular quadrilateral: has four equal sides and four equal angles

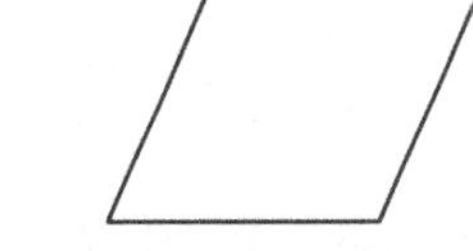

Not a regular quadrilateral: has four equal sides, but angles are not all equal

There are several other quadrilaterals with special names.

A rectangle is a simple quadrilateral with four right angles. A square is a simple quadrilateral with four equal sides and four equal (right) angles. Yes, a square is the regular quadrilateral you were looking for a couple of paragraphs ago. A square is a rectangle, but a rectangle might not be a square. Why?

A kite is a simple quadrilateral with two pairs of adjacent sides the same length.

A pair of sides are parallel if they are contained in a pair of parallel lines. That is, if the lines created by extending two sides of a quadrilateral forever in both directions never touch, the two sides are parallel sides. A trapezoid is a simple quadrilateral with only one pair of parallel sides [25]. If you already know what a trapezoid is, you are probably picturing an isosceles trapezoid. An isosceles trapezoid is a trapezoid with its two nonparallel sides the same length.

A parallelogram is a quadrilateral with two pairs of parallel sides.

See the examples of special quadrilaterals below.

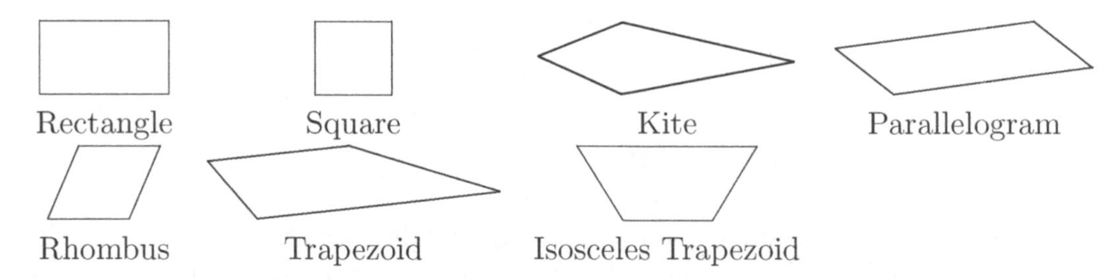

Understanding the Text:

1. Consider the four quadrilaterals below. Classify each of the quadrilaterals as convex or simple. Some will be both and some will be neither.

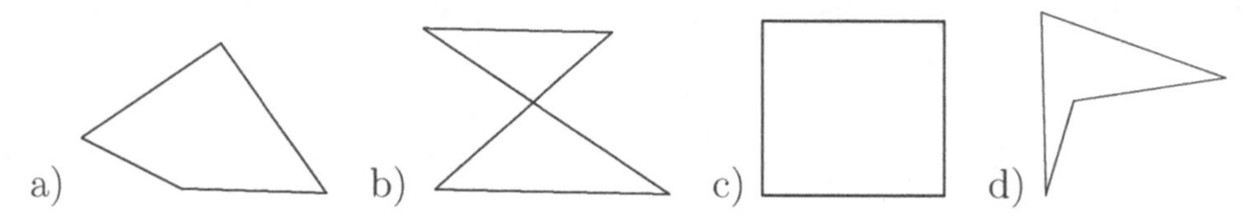

2. Draw an example of a quadrilateral that is convex but not simple, or explain why there cannot be such a quadrilateral.

3. Draw an example of a quadrilateral that has four equal angles but not four equal sides, or explain why there cannot be such a quadrilateral.

4. Is a square a rhombus? Explain your answer.

5. Is a parallelogram a trapezoid? Explain your answer.

6. Is a rhombus a square? Explain your answer.

7. Is a square an isosceles trapezoid? Explain your answer.

8. Is a trapezoid a parallelogram? Explain your answer.

4.2 Special Angles of Quadrilaterals

Quadrilaterals have the same types of special angles that triangles do: central angles, vertex angles, and exterior angles. Just as with triangles, a central angle is an angle whose vertex is the center of the quadrilateral with rays that intersect adjacent vertices of the quadrilateral. A vertex angle is an angle whose vertex is the vertex of a quadrilateral with rays containing the edges emanating from (coming out of) that vertex. An exterior angle is an angle that makes a straight line with the vertex angle. See the examples below.

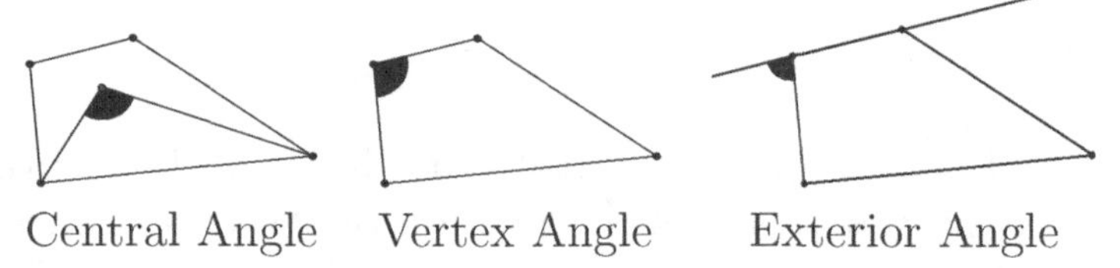

Central Angle Vertex Angle Exterior Angle

What are the special angle measures of a regular quadrilateral (square)? The central angles of a regular quadrilateral are all the same, and if you put them all together, they go all the way around the center of the regular quadrilateral. See the figures below. To go all the way around in a circle is 360°; if we divide 360° up into four angles the same size, we get $\frac{360}{4} = 90$. Therefore, the measure of each central angle in a regular quadrilateral is 90 degrees.

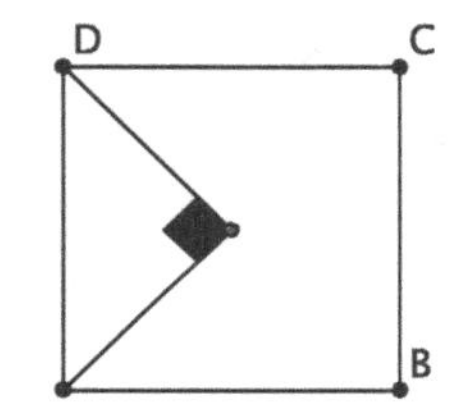

One Central Angle of a Square

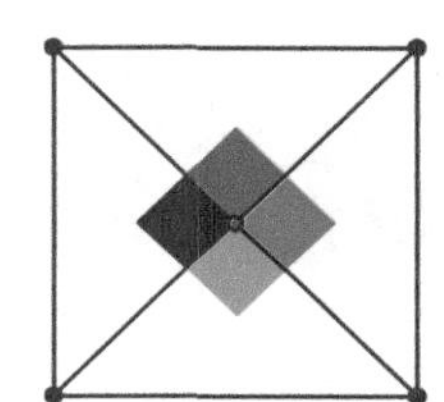

All Central Angles of a Square

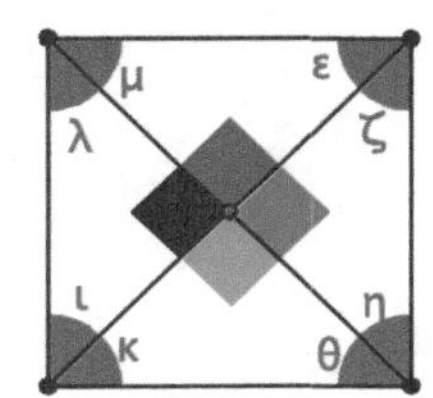

Angles of Triangles made by Central Angles

The rays of the central angles of a regular quadrilateral create four triangles in the square when they cross the vertices of the square. Look at the last picture in the figure above for an illustration of this. We would now like to compute the measure of the vertex angles of a regular quadrilateral. You probably know already what we should get for an answer, so let us see if we can reason it out.

Recall that the sum of the angles of any triangle is 180°. Each of these four triangles has one angle that measures 90°. Therefore, the sum of the other two angles of each of these triangles must be $180° - 90° = 90°$. The sum of the measure of the remaining angles is the sum of all of the vertex angles of the square. Since there are four triangles, this means that the sum of the measures of all of the vertex angles of the regular quadrilateral must be $90 \times 4 = 360°$. Since there are four angles in a regular quadrilateral and each of the angles is the same size (otherwise, the figure would not be regular), each angle of the regular quadrilateral must have 90°. Is that what you thought the answer should be?

What are the measures of the exterior angles of a regular quadrilateral? The exterior angles of any quadrilateral make a straight line with the vertex angles (just like they do in a triangle). Therefore, the measure of any exterior angle of a regular quadrilateral is $180° - 90° = 90°$.

Understanding the Text:

1. Draw a simple quadrilateral on a piece of paper and cut it out. Shade all four corners of the quadrilateral. Cut off the corners so that only a portion of the cut-off piece is shaded. Take the pieces you cut off of your quadrilateral and lay them on a hard, flat surface (such as a table) edge to edge, with the shaded corners touching each other. If you combine all four angles, what angle do they make? If you added the measures of all of the angles of your quadrilateral together, what would you get?

Chapter 5

Polygons and Special Angles

5.1 Polygons

In this chapter, we are going to put together and expand upon the chapters on Triangles (Chapter 3), and Quadrilaterals (Chapter 4). Triangles and quadrilaterals are, in fact, types of polygons. The word "poly" means "many" and "gon" comes from the Greek word "gōníō" which means "angle."[27] Therefore, polygon means "many angles." A polygon is a figure created by connecting n noncollinear vertices with n edges so that each vertex is connected to exactly two edges, and each edge is connected to exactly two vertices (one at each end). Noncollinear vertices means that there is no line that contains three of the vertices at the same time.

Just as with quadrilaterals in Section 4.1, there are simple, complex, and convex polygons. The definitions of these terms for polygons are analogous to their definitions for quadrilaterals.

A simple polygon is a polygon with edges that intersect only at their endpoints. In a simple polygon, the edges do not cross each other. See the example of a simple polygon and an example of a polygon that is not simple below.

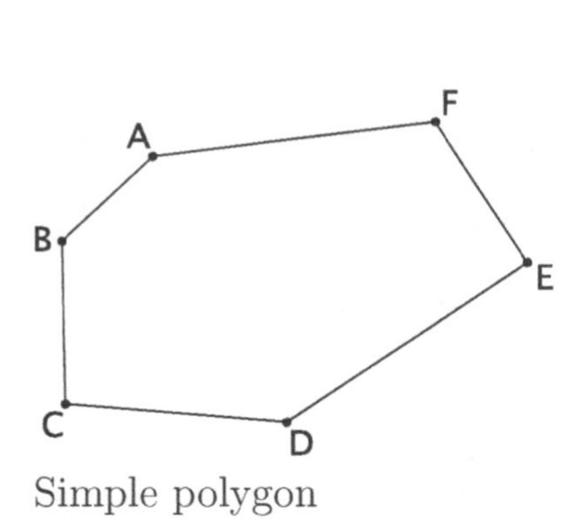

Simple polygon

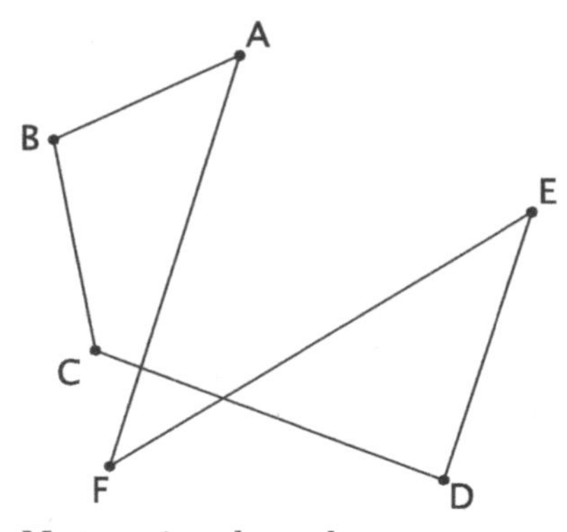

Not a simple polygon:
the edges intersect.

A simple polygon has an "inside" and an "outside" of the polygon. Since the sides do not cross each other, they create a "fence" that contains a space called the "inside" of the quadrilateral. If a polygon is not a simple polygon, it is called a complex polygon.

A convex polygon is a simple polygon for which any line segment between two points in or on the polygon stays completely within that polygon. The following polygon is a convex polygon.

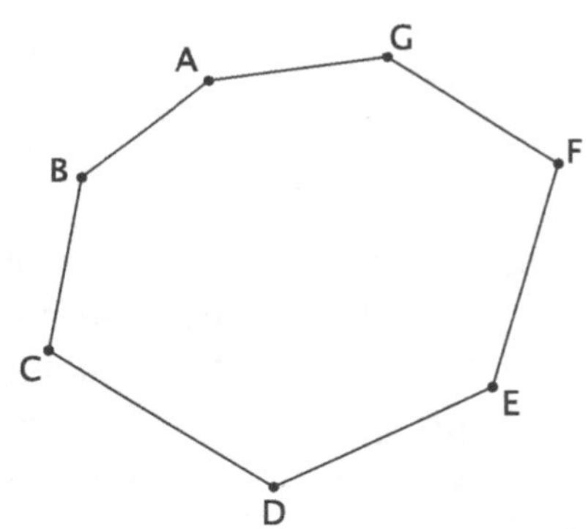

The following polygon is not a convex polygon:

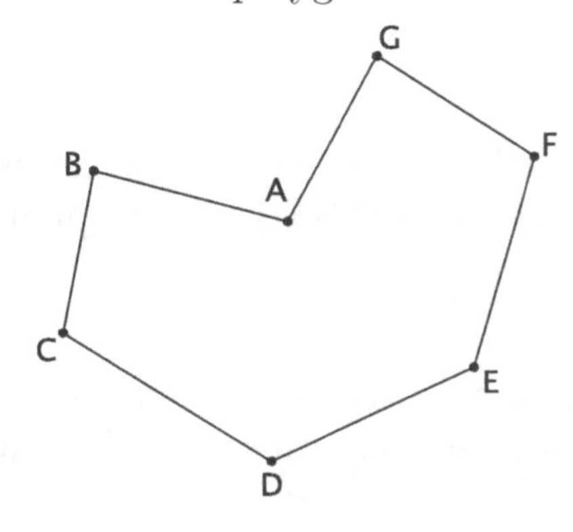

This is because there is at least one line segment between two points of the polygon that does not stay completely within the polygon. The line segment $\overline{HI}$ below has endpoints H and I, each of which are inside the polygon, but the point J on the line segment is one example of a point on the line segment $\overline{HI}$ that lies outside of the polygon.

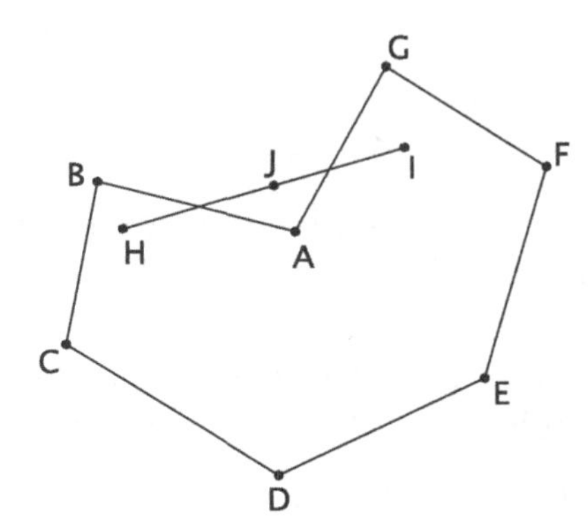

A regular polygon is a convex polygon in which all sides are of the same length and all vertex angles of the polygon are of the same size. If a convex polygon has all sides with the same length, it does not necessarily follow that the vertex angles are of the same size.

Polygons have special names based on the number of sides they have:

Number of Sides	Name
3	Triangle
4	Quadrilateral
5	Pentagon
6	Hexagon
7	Heptagon
8	Octagon
9	Nonagon
10	Decagon
12	Dodecagon

A regular polygon is a polygon with all sides of the same length and all angles of the same size. A regular triangle is called an "equilateral triangle," and a regular quadrilateral is called a "square." None of the other regular polygons has a special name. They are just called "regular" whatever it is: regular hexagon, regular decagon, etc.

Understanding the Text:

1. Is it possible for a polygon to be both simple and convex? If so, give an example of such a polygon. If not, explain why.

2. Is it possible for a polygon to be both complex and convex? If so, give an example of such a polygon. If not, explain why.

3. Is it possible for a polygon to be simple but not convex? If so, give an example of such a polygon. If not, explain why.

5.2 Special Angles of Polygons

All polygons have central angles, vertex angles, and exterior angles. Their definitions are the same as they were for quadrilaterals in Section 4.2. The method for finding them for a regular polygon is also the same.

For example, for a regular pentagon, the central angle is a full circle (360°) divided evenly among the five central angles in a regular pentagon. Therefore, the central angle of a pentagon is $\frac{360°}{5} = 72°$. See the figure below for a picture of all of the central angles of a regular pentagon.

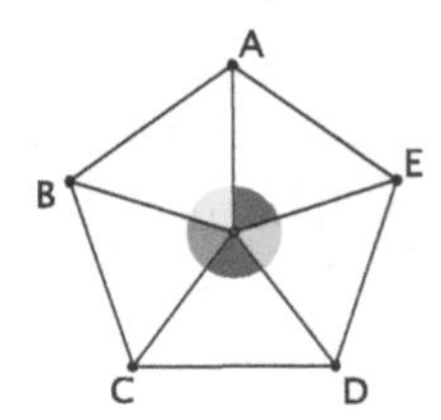

Now, to find the vertex angles, consider the triangles made when the rays of the central angles intersect the vertices of the regular pentagon.

As with the computation of the vertex angles of a square, we can use the fact that the sum of all the angles of any triangle is 180°. The rays of the central angles along with the edges of a regular pentagon make five identical triangles in a pentagon. If we take the measure of each central angle out of each triangle, we get $180° - 72° = 108°$. If we put all of the remaining vertices of these triangles together, we get all of the vertex angles of the regular pentagon. Therefore, the sum of the measures of the vertex angles of a regular pentagon is

$$108° * 5 = 540°$$

Then the measure of each vertex angle is

$$\frac{108° * 5}{5} = \frac{540°}{5} = 108°$$

The exterior angle is an angle that makes a straight line with one of the vertex angles of the regular polygon. In this case, that means that the measure of the exterior angle is $180° - 108° = 72°$.

One thing I notice about these numbers is that the measure of the exterior angle and the measure of the central angle are the same. Is this a coincidence? In fact, it is not. Let us go back to the five triangles made when the rays of the central angles meet the vertices of the regular hexagon. Since the polygon is a regular polygon, the rays of the central angles coming out of the center of the regular pentagon are of the same length. That means that each of those triangles are isosceles triangles. One thing we learned at the very end of Section 3.1 is that if two sides of a triangle are of the same length, then the angles opposite those sides will be of the same size.

Therefore, each of the angles of the triangles that are not central angles of the pentagon have measure

$$\frac{180^\circ - \dfrac{360^\circ}{5}}{2} = 54^\circ$$

Two of these angles together make one vertex angle of a regular pentagon. Therefore, the vertex angle of a regular pentagon is

$$\frac{180^\circ - \dfrac{360^\circ}{5}}{2} * 2 = 54^\circ * 2$$

$$180^\circ - \frac{360}{5} = 108^\circ$$

So, in fact, it is the case that the measure of the central angle (in this case, $\frac{360^\circ}{5} = 72^\circ$) is supplementary to the vertex angle of the regular pentagon. Since both the exterior angle of a regular pentagon and the central angle of a regular pentagon are supplementary to the vertex angle of a regular pentagon, they are equal to each other.

Understanding the Text:

1. Find the central angle and vertex angle of a regular hexagon.
2. Find the central angle and vertex angle of a regular heptagon.
3. Find the central angle and vertex angle of a regular octagon.
4. Find a formula to find the central angle and vertex angle of any regular polygon. Use n for the number of sides of your polygon.

Chapter 6

Symmetry and Isometry in Two-Dimensions

6.1 Symmetry of Polygons and Other Figures

There are two different types of symmetry of a particular object: rotational symmetry and reflectional symmetry. Rotational symmetry is the type of symmetry you find in a pinwheel:

A pinwheel has a center that the whole object rotates around, and at particular angles, the pinwheel looks the same as it did when you started. Ignoring color in the pinwheel above, there are eight places where the figure looks the same. Notice that if you spin it all the way around, you have spun everything around a circle. Since an entire circle is 360°, each place where the pinwheel looks the same (again, ignoring color) is $\frac{360°}{8} = 45°$. Therefore, the rotational symmetry of this pinwheel is each 45° turn around the center of the pinwheel; I will call this the "interval." We write down the rotational symmetry by specifying each angle, starting with the original position, in which the pinwheel looks the same as when we started and by specifying the center of the rotation. Therefore, the rotational symmetry of the pinwheel is 0°, 45°, 90°, 135°, 180°, 225°, 270°, and 315° around the center of the pinwheel.

Since there are eight places where the pinwheel looks the same, there should be eight angles in the list. Also, 0° is the original position; if you do not move the pinwheel at all, it looks the same as when you started. If you add another 45° to 315°, you get 360°, which puts you back in the original position. That makes sense, as you should have gone full circle at that point. If you do not get back to 360° after adding the interval once more, one of several things may have happened; 1) you added incorrectly; 2) you rounded your answers, so the last step does not quite come out even; 3) You have not added the interval enough times; 4) You were wrong about the interval in the first place.

As in the discussion in the previous paragraph, in order to specify the rotational symmetry of any object, you need to specify the center of rotation and the degrees of rotation that leaves the object looking identical to how it looked when you started.

Let us look at another example of rotational symmetry—a regular octagon.

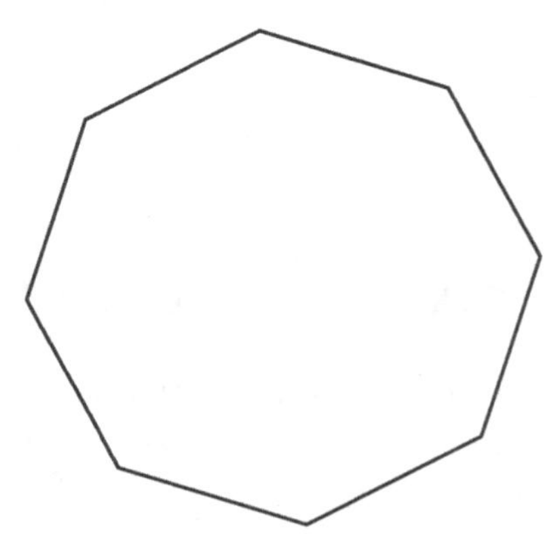

There are eight places where a regular octagon looks the same if you turn it around its center. Therefore, just as with the computation of a central angle, the angle of rotation from the original position to the next position where the octagon looks the same is $\frac{360°}{8} = 45°$. See the picture below.

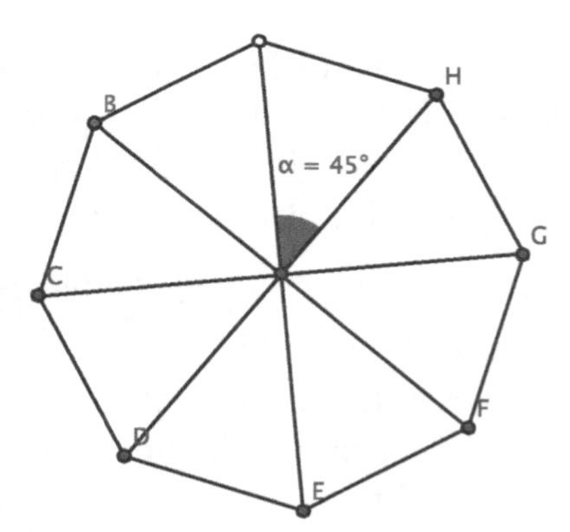

If asked to show the rotational symmetry of an object, you need to show the center of rotational symmetry and all of the angles of symmetry for that object.

Therefore, the answer to the question is demonstrated like this:

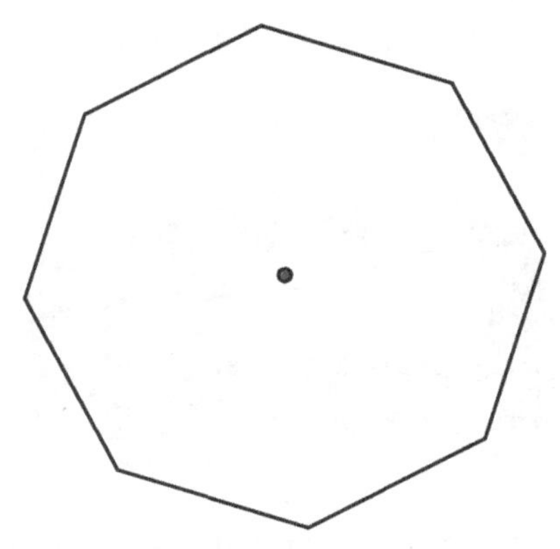

0°, 45°, 90°, 135°, 180°, 225°, 270°, 315°

Notice again that if you add another 45° to 315°, you get 360°, which is what we expect. The reason we use 0° in the list rather than 360° is because 0° is the "identity element" in symmetry. Just like the number 0 is the identity element in addition and the number 1 is the identity element in multiplication, the 0° rotation is the identity rotation in symmetry movements. For more information on this, see Section 6.2.

For an example of a rotation that is not a symmetry, take a square and rotate it 45°. In the pictures below, the figure on the left is the original square. The figure on the right is the original square along with a rotation of that square counterclockwise an angle of 45°. As you can see, this particular rotation does not land precisely in the footprint of the original figure. In this case, the vertices of the square land in a place where an edge of the original square was located.

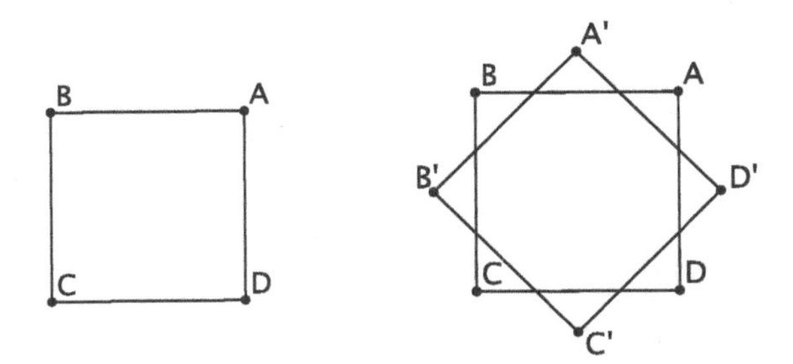

The other type of symmetry of an object is reflectional symmetry. To specify reflectional symmetry, we have to specify the lines of symmetry. A line of symmetry of an object is a line through the object which divides the object in half so that one half is the mirror image of the other half. With two-dimensional objects, you can think of the line of symmetry as a fold line; if you fold the object on the line of symmetry, the two sides should match up exactly.

An example of an object with reflectional symmetry is the house below. The line of symmetry runs down the center of the house. In this case, when looking for symmetry, disregard the door, and focus on the main structure of the house.

©alexmisu/Shutterstock.com

Show reflectional symmetry by showing the lines of symmetry in the object, as shown below.

©alexmisu/Shutterstock.com edited by Amy Wangsness Wehe

Now we know how to find the reflectional and rotational symmetry of polygons and other shapes with corners. What about a circle?

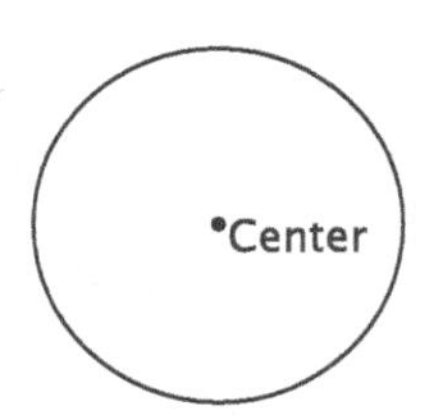

If you rotate a circle around its center, it looks the same no matter how many degrees you turn it. This holds for 15° and 24° as well as 2.456° and 359.875°. That means it has rotational symmetry about its center for any number of degrees from 0° to 360°. Since there is not a number of degrees "right before" 360° in this case, we have to write the 360°; technically, we would write this interval as $[0°, 360°)$, where the "[" means "including 0°" and the ")" means "not including 360°."

There are also infinitely many lines of reflectional symmetry. One such line is shown in the picture of the circle below. A line of reflectional symmetry of a circle must go through the center of the circle, but that is the only requirement. Any line through the center of a circle will divide the circle into two equal parts. It is, therefore, a line of reflectional symmetry of the circle.

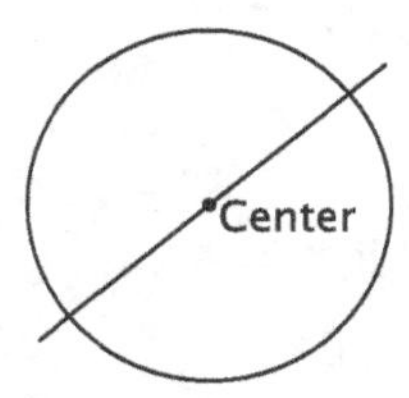

Some other objects also have both rotational and reflectional symmetry. The honeysuckle flower, ignoring the stamen, or center part, of the flower has both types of symmetry. A honeysuckle flower [12] is pictured below.

For instance, a honeysuckle flower has five petals and five round parts between the petals. If you were to draw the outline of the flower on paper and then turn the flower so that one of the petals lands in the outline of where an original petal was, the whole flower will fill in the original outline. At what angle do you need to turn the flower so that it looks exactly the same again? A full circle is 360° and there are five positions in which the flower looks the same. The petals are equally spaced, so that means there are $\frac{360}{5} = 72°$ between them. Therefore, the honeysuckle flower has $0°, 72°, 144°, 216°$, and $288°$ rotational symmetry around the center of the flower.

One example of a line of symmetry of the honeysuckle flower is given in the picture below. If everything were lined up perfectly, there would be a line of reflection through the tip of each pedal and the center of the flower. This line would also go through the middle of one of the rounded pieces of the flower and between two other rounded pieces of the flower. Therefore, a honeysuckle flower arranged perfectly symmetrically will have five lines of reflectional symmetry.

Consider the regular nonagon (nine-sided regular polygon) below:

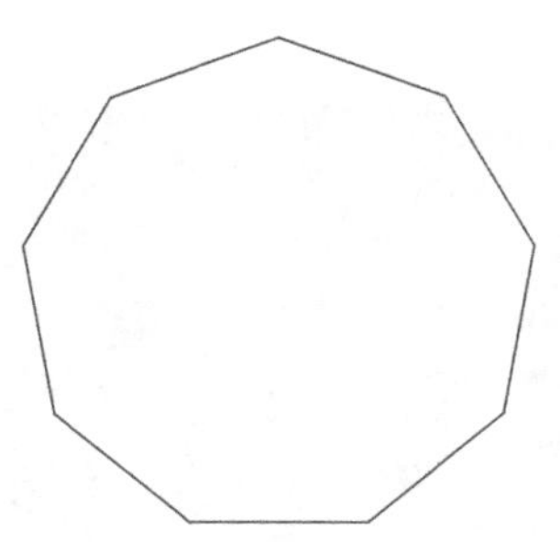

To find the rotational symmetries of the polygon, we consider where each part of the nonagon could land. A vertex, for instance, would have to land on another vertex in order for the nonagon to land in its own footprint. Therefore, there are only nine places it could land when rotated around the center of the polygon. Likewise, there are nine edges, and each edge can only land on another edge. Since these are the only components of the polygon, we now know that there are nine places that the nonagon looks the same when rotated around its center. Therefore, the rotational symmetry of the nonagon is given in the picture below:

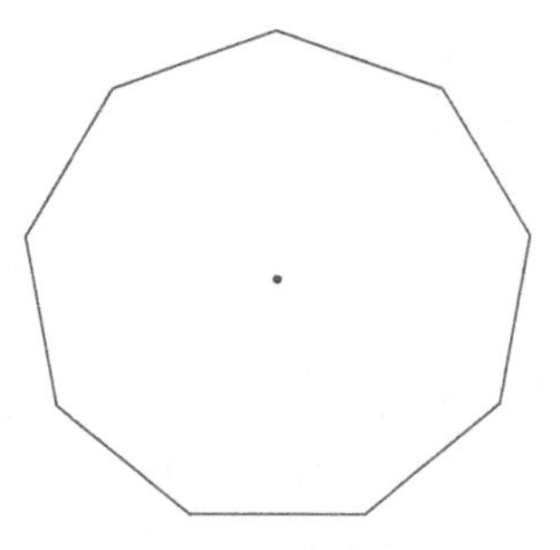

$0°, 40°, 80°, 120°, 160°, 200°, 240°, 280°, 320°$

To find the reflectional symmetries of the nonagon, we look for lines through the figure where the object looks the same on both sides of the line. In this case, we see that if we draw a line from the vertex of the nonagon to the midpoint of the opposite side of the nonagon, the figure looks the same on both sides of the line. This is the case for any such line through the nonagon. Therefore, there is a line for each vertex of the nonagon, and the picture below illustrates the lines of symmetry of the figure.

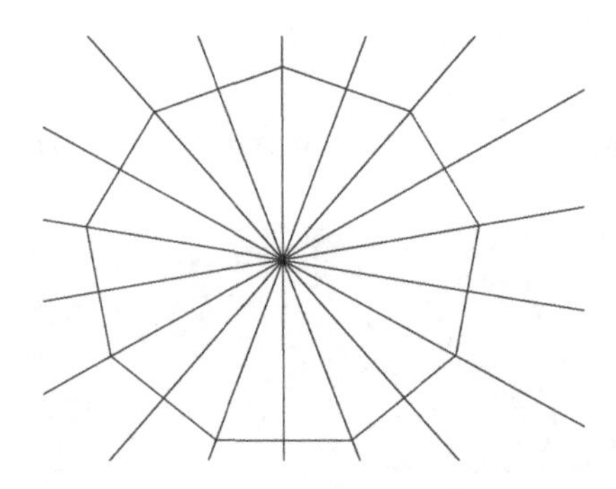

Designs within a figure can decrease the number of symmetries it has. For instance, a regular octagon has eight rotational symmetries, as shown below:

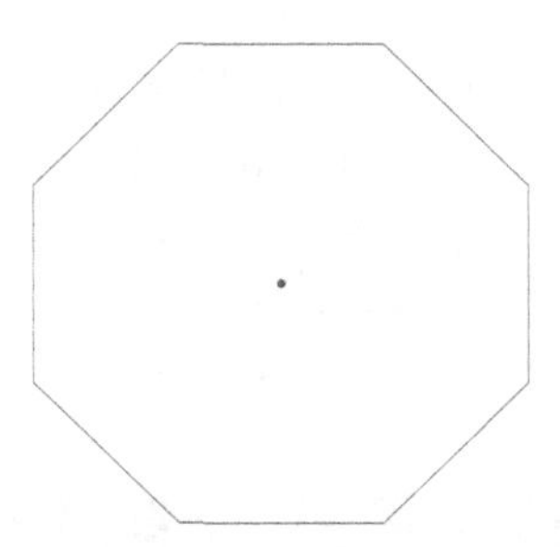

$0°, 45°, 90°, 135°, 180°, 225°, 270°, 315°$

If a square is added to the interior of the regular octagon, however, then not only must the vertices and edges of the octagon land on their corresponding parts, but the edges of the square must also land on edges of the square. This decreases the number of places the figure looks the same when you turn it. In this case, there are now only four places where the figure looks the same, as shown in the picture on the following page:

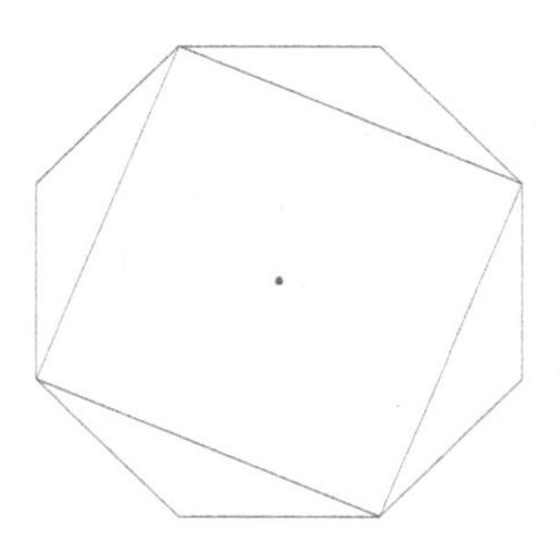

$0°, 45°, 90°, 135°, 180°, 225°, 270°, 315°$

The purple flower [20] below has neither rotational nor reflectional symmetry.

Understanding the Text:

1. Show the symmetries of the image of the Norwegian flag below [19]. There are two copies of the object; show the rotational symmetries on one and the reflectional symmetries of the object on the other. If the object does not have a certain type of symmetry, state "there is no rotational/reflectional symmetry" next to the image. When finding the symmetries, go with the general shape of the object and ignore slight imperfections.

2. Show the symmetries of the image of the regular heptagon below. There are two copies of the object; show the rotational symmetries on one and the reflectional symmetries of the object on the other. If the object does not have a certain type of symmetry, state "there is no rotational/reflectional symmetry" next to the image.

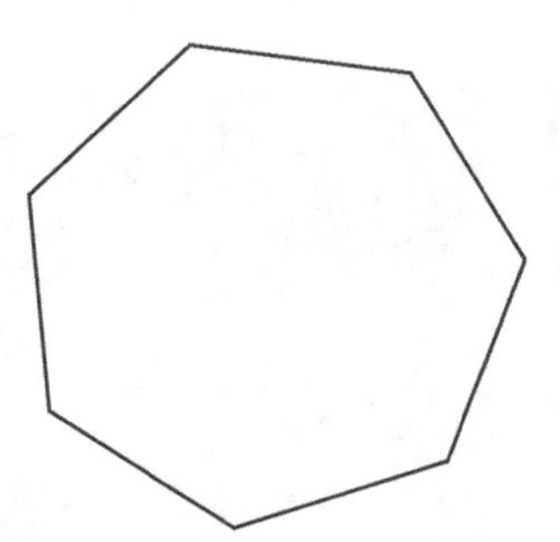 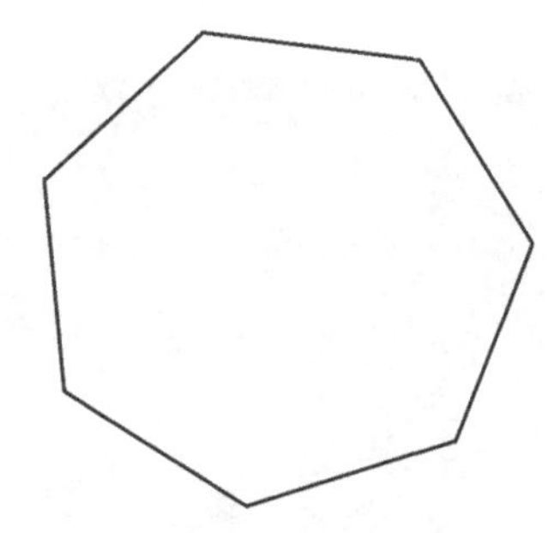

3. Show the symmetries of the image of the pinwheel below. When looking for symmetries, disregard the stick that the pinwheel is connected to. There are two copies of the object; show the rotational symmetries on one and the reflectional symmetries of the object on the other. If the object does not have a certain type of symmetry, state "there is no rotational/reflectional symmetry" next to the image.

©opicobello/Shutterstock.com edited by Amy Wangsness Wehe

4. Show the symmetries of the image of the lily below [10]. There are two copies of the object; show the rotational symmetries on one and the reflectional symmetries of the object on the other. If the object does not have a certain type of symmetry, state "there is no rotational/reflectional symmetry" next to the image. When finding the symmetries, go with the general shape of the flower and ignore slight imperfections.

5. What regular polygon has the same rotational and reflectional symmetries as the picture in Problem 4?

6. Show the symmetries of the image of the Buttercup below [4]. There are two copies of the object; show the rotational symmetries on one and the reflectional symmetries of the object on the other. If the object does not have a certain type of symmetry, state "there is no rotational/reflectional symmetry" next to the image. When finding the symmetries, go with the general shape of the flower and ignore slight imperfections.

7. What regular polygon has the same rotational and reflectional symmetries as the picture in Problem 6?

8. Show the symmetries of the image below. There are two copies of the object; show the rotational symmetries on one and the reflectional symmetries of the object on the other. If the object does not have a certain type of symmetry, state "there is no rotational/reflectional symmetry" next to the image.

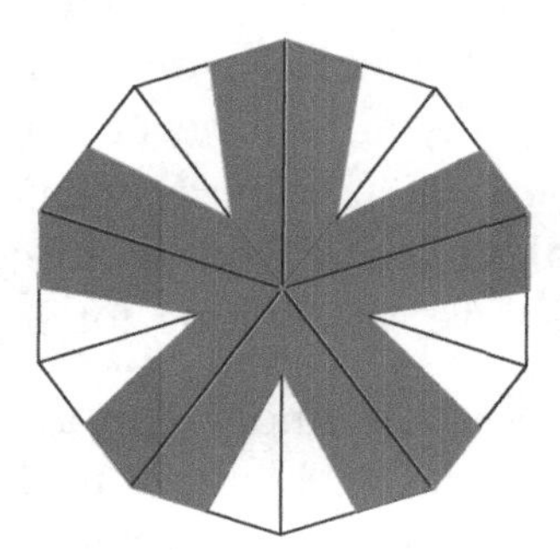 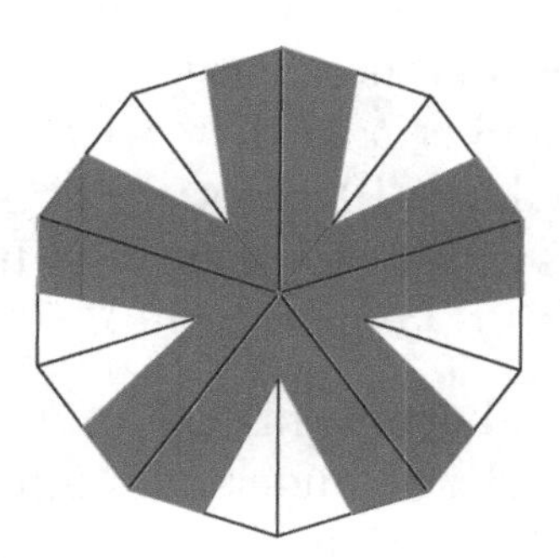

6.2 Symmetry Groups

A group is used in mathematics to generalize sets and operations that we use every day. One example of a group is the integers under the operation addition $+$. Integers are the positive and negative whole numbers plus zero, that is, they are

$$\cdots - 5, -4, -3, -2, -1, 0, 1, 2, 3, 4, 5, \ldots$$

The example of a group in the paragraph above consists of a set and operation that we see every day. What makes the integers under the operation addition a group? We first state the set we are working with (integers) and the operation we are using (addition).

In order to be a group, all of the following four items must hold:

1. The set must be **closed** under the operation. This means that if you add any two numbers in the set, the answer is also part of the set. If you take two integers and add them together, do you get another integer? The answer is yes.

2. The set must contain an **identity element**. This means that there must be a number which, when added to any other number, does not change that number. That is, we need a number x for which $2+x=2$, $x+10=10$, $5+x=5$, and so on. What number is that? The number we are looking for is zero.

3. The operation must be **associative** under the set. That is, given any three numbers in the set a, b, c, not necessarily unique (not necessarily different from each other), $a + (b + c) = (a + b) + c$. You probably learned long ago that addition is associative. That is, you can add the last two numbers before adding the first and get the same answer as if you add the first two numbers and then add the third one to it.

4. Finally, every element in the set must have an **inverse** element. That means any number in the set must have another number in the set such that the sum of these two numbers is the identity. In this case, that means for any number a, there must be another number b so that $a + b = 0$. If I give you an integer, say -10, what number, when added to -10, gives you 0? If I give you the number 5, what number, when added to 5, gives you the number 0? In general, if I give you an integer a, what number, when added to a, gives you 0? The number we are looking for is $-a$.

If all of the above items hold, the set is a group under the given operation. Since the integers are closed under addition, they contain an identity under addition, they are associative under addition, and every integer has an inverse under addition, the integers are a group under addition.

If the numbers in the set can also be added in any order, the group is called a commutative group. You know from previous experience that this is true of integers. For instance, $2 + 4 = 4 + 2$ and $100 + 24 = 24 + 100$.

Another example of a group is the group of the symmetry movements of a square under the operation "followed by." This is a group we do not think about every day. Consider a square. If you do not rotate the square at all (leave it alone), it looks the same as it did to begin with. If you rotate it 90° counterclockwise, it looks the same as it did to begin with. You can also rotate it 180° and 270° counterclockwise, and it will still look the same as when it started. You can also take the square and reflect it over a horizontal line dividing it in half, and it will still look the same. Likewise, you can reflect a square over a vertical line dividing it in half or over either of its diagonals, and it will look the same as when you started. The figure below and on the following page demonstrate all of these movements. In mathematics, the standard direction for rotation is a counterclockwise rotation; therefore, all of the rotation movements in the figures are in the counterclockwise direction.

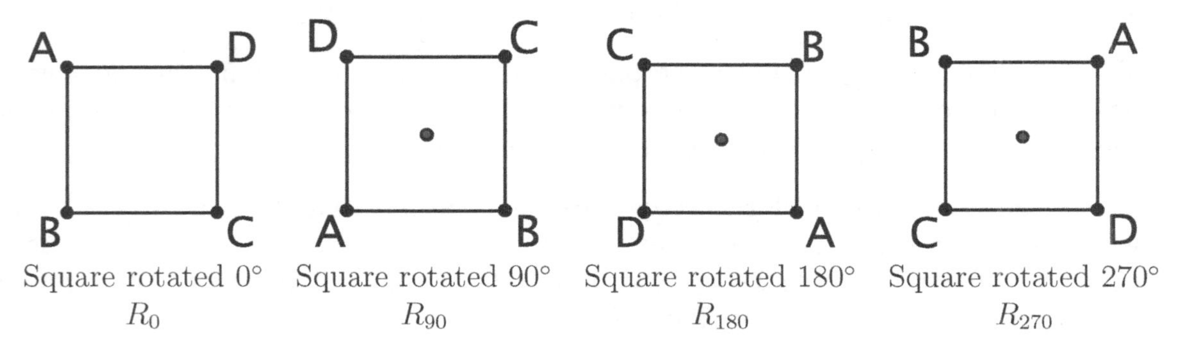

Square rotated 0° R_0 | Square rotated 90° R_{90} | Square rotated 180° R_{180} | Square rotated 270° R_{270}

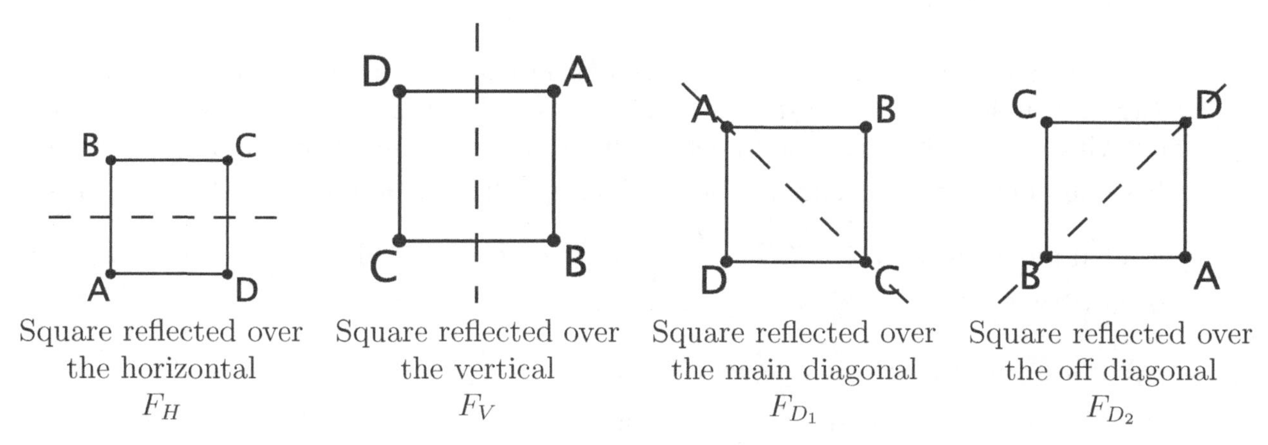

Square reflected over the horizontal F_H

Square reflected over the vertical F_V

Square reflected over the main diagonal F_{D_1}

Square reflected over the off diagonal F_{D_2}

The operation "followed by" means to take one symmetry movement followed by another movement. For example, if you take the original square and rotate it counterclockwise 180° followed by reflecting it over the vertical symmetry line, it would be the same as starting with the original figure and just reflecting the square over the horizontal symmetry line. See the figure below for an illustration. Pay particular attention to the labels of the vertices, since the squares themselves all look the same.

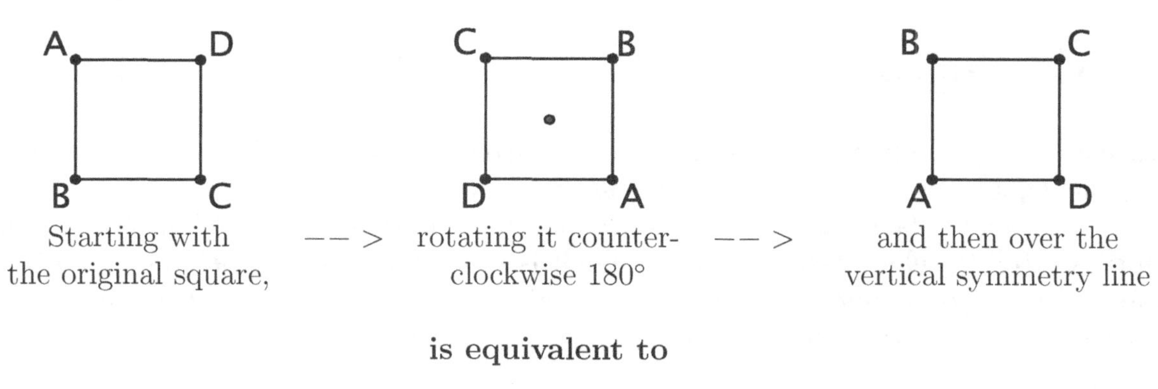

Starting with the original square, $-->$ rotating it counter-clockwise 180° $-->$ and then over the vertical symmetry line

is equivalent to

starting with the original square $-->$ and reflecting it over the horizontal symmetry line

We write this operation as $R_{180} \cdot F_V = F_H$ where the "$\cdot$" represents the "followed by" operation. The notation for the rotation and the flips (R_{180}, F_V, and F_H) are listed in the figure of symmetry movements at the top of this page and the previous one.

Understanding the Text:

1. Another example of a group is the rational numbers excluding zero under the operation multiplication. The rational numbers are the fractions, excluding any fractions involving the number zero (no undefined fractions and no fractions equivalent to zero); these numbers are any numbers of the form $\frac{a}{b}$ where $a \neq 0$ and $b \neq 0$. Examples of rational numbers are $1, 5, -7, \frac{1}{2}, \frac{-4}{7}, \frac{10}{3}$, and so on. State all of the elements necessary to show that this is a group. In particular, answer the following:

   (a) State the set and the operation in question.

   (b) Is the set closed under the operation?

   (c) Does the set have an identity element under the operation? If so, what is it?

   (d) Is the set associative under the operation?

   (e) Does every element of the set have an inverse under the operation?
   If all of the above questions can be answered in the affirmative, then the rational numbers excluding zero under the operation multiplication are a group.

6.3 Tessellations

A tessellation or tiling is a covering of an infinite geometric plane without gaps or overlaps by congruent plane figures of one type or a few types [30].

The "congruent plane figures" mentioned in the definition of a tessellation are called "prototiles." If a tessellation has one prototile, it is called monohedral. See the figure below for a few examples of monohedral tessellations. What you see is only part of the tessellation; these patterns go on forever in all directions.

Tessellation of a Square

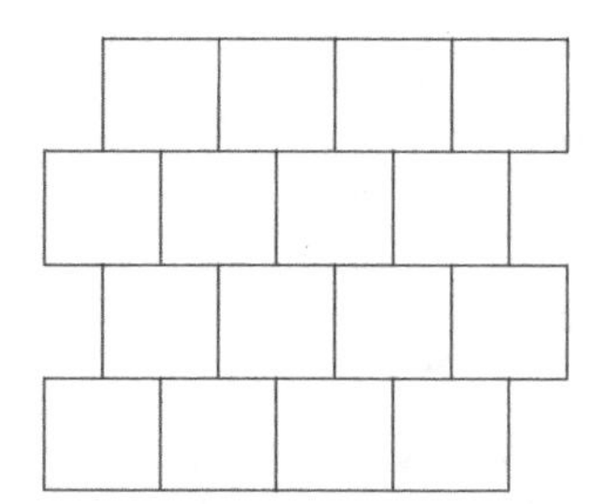
Tessellation of a Square

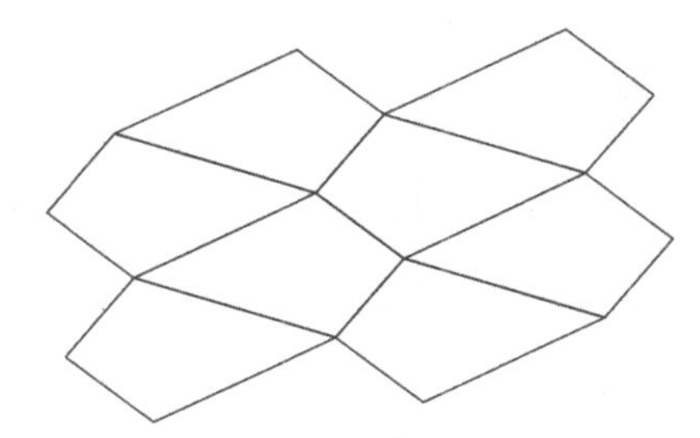
Tessellation of a Quadrilateral

A tessellation that has two prototiles is called dihedral. Below are some examples of dihedral tessellations. Again, since tessellations go on forever in all directions, what you see is actually only part of the entire tessellation.

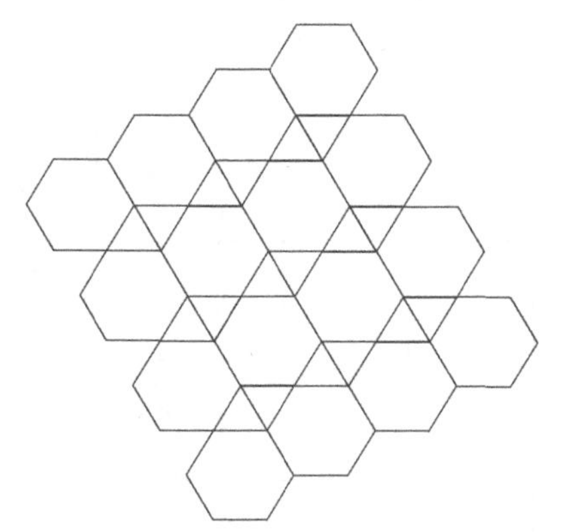

Tessellation of a Regular Hexagon and an Equilateral Triangle

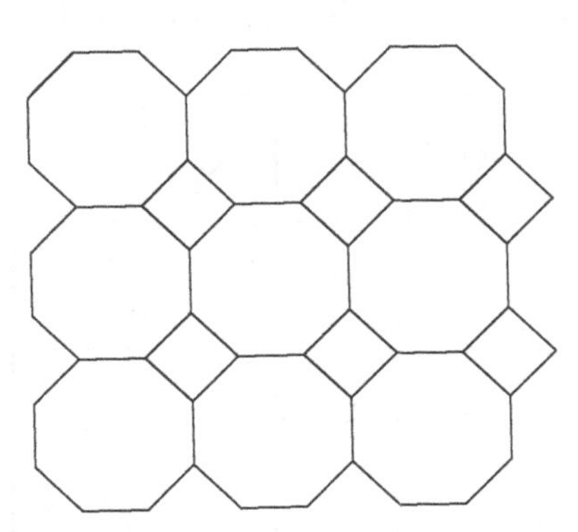

Tessellation of a Square and a Regular Octagon

A tessellation that has three prototiles is called trihedral. After that, we just say k-hedral, as in 4-hedral, 5-hedral, and so on.

In a tessellation, an edge is defined to be the boundary between two prototiles. In the first tessellation in the figure below, the edges of the tessellation are also the edges of the prototiles; one such edge is bolded. In the second tessellation in the figure below, the edges of the tessellation are not the same thing as the edges of the prototile; again, one edge of the tessellation is bolded. In each case, the edge is the boundary between the two prototiles marked with an "x."

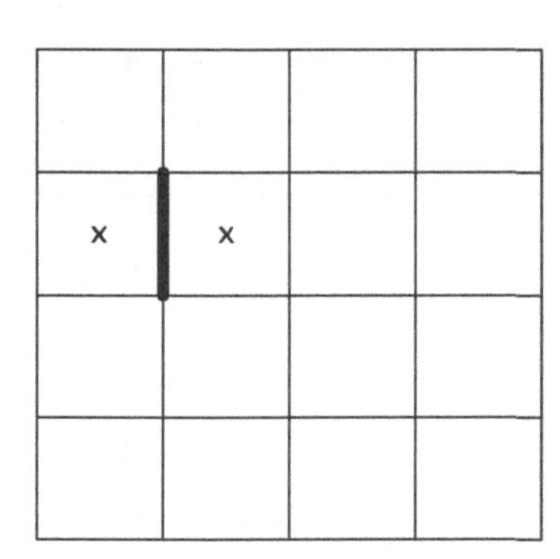

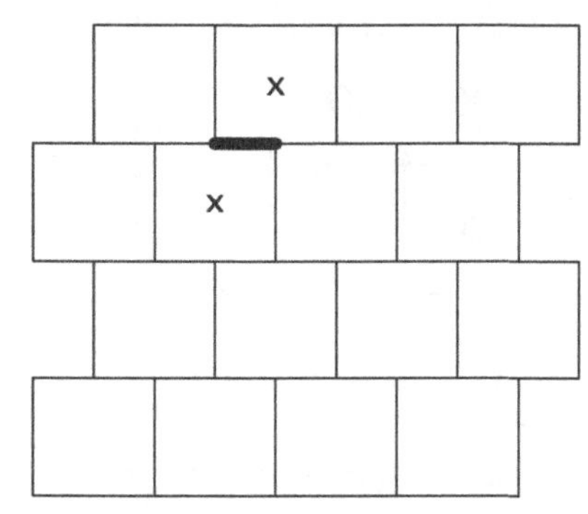

A vertex of a tessellation is an endpoint of an edge of the tessellation.

In a tessellation, two prototiles are neighbors if they share at least one point in common. Any particular prototile in the figure below has eight neighbors. To see this, choose a square that is not on the edge of the picture below. There are four squares that share an edge and four squares that share a corner (vertex) with the square you chose. Since the corners are also points, those squares also share a point with the square you chose. Therefore, there are eight neighbors of the square you choose. In the figure on the next page, the square I chose has an "X" in the middle of it. The squares that are neighbors to that square have an "O" in them. Note that in this tessellation, the only prototile is a square.

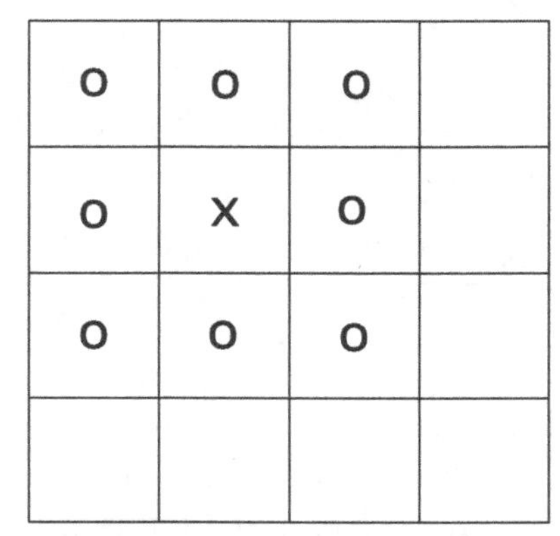

The "X" is the chosen square and the "O"'s are the neighbors

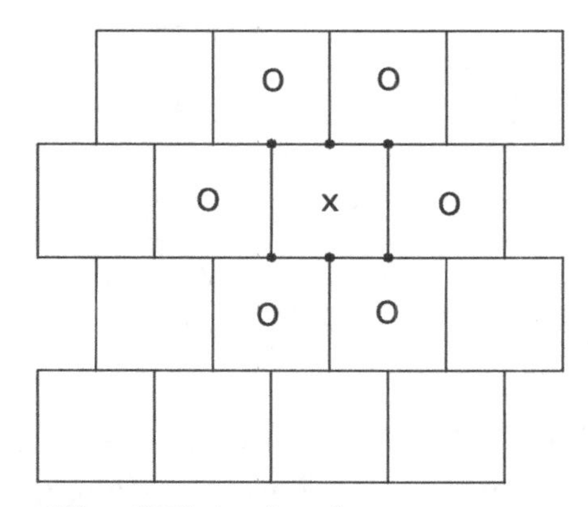

The "X" is the chosen square and the "O"'s are the neighbors

Two prototiles are adjacent if they share an edge. In the figure below, the prototile that has an "X" in it is the chosen prototile. The tiles with an "O" in them are adjacent to "X". The endpoints of the edges they share are shown; these are some of the vertices of the tessellation.

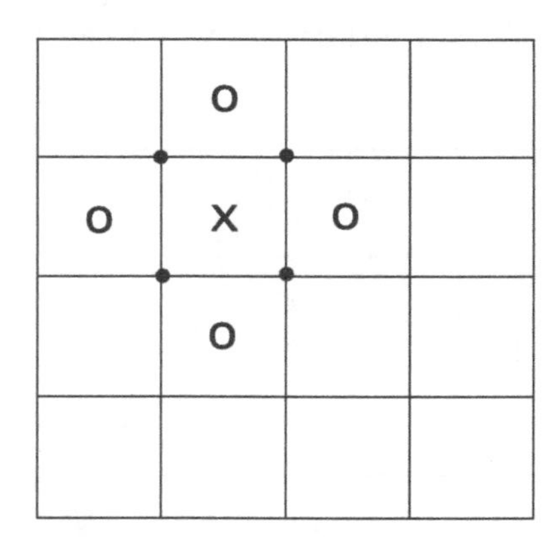

The "X" is the chosen square and the "O"'s are adjacent to "X"

The "X" is the chosen square and the "O"'s are adjacent to "X"

The vertex figure of a tessellation is a polygon whose vertices are the midpoints of the polygonal edges emanating (coming out of) a vertex of the tessellation. The best way to read this definition is to read it backward. First, find a vertex of the tessellation; chose a vertex that is surrounded by edges and not one on the edge of the picture. Then, find all of the edges coming out of that vertex. Next, draw in the midpoints of all of those edges. Finally, connect all of those midpoints. The resulting polygon is a vertex figure of the tessellation. The picture below illustrates the vertex figure of a tessellation we have already seen. Vertex V is the chosen vertex, the other four points shown are the midpoints of the emanating edges.

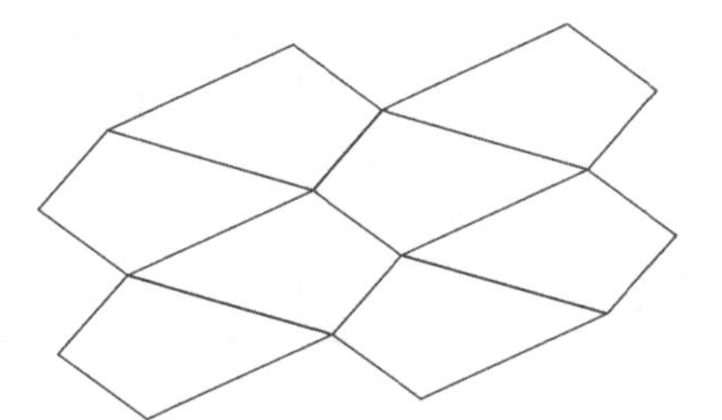
The Original Tessellation

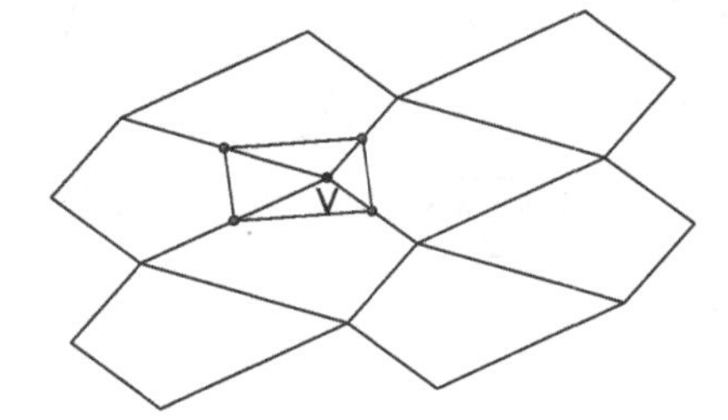

Tessellation with Vertex Figure

A tessellation is called a regular tessellation if all of the prototiles are convex regular polygons of the same size and shape and if all of the vertex figures are also regular polygons.

There are only three polygons that make regular tessellations: equilateral triangles, squares, and regular hexagons. All other monohedral tessellations are made with figures that are not regular.

Why is this? First, since there is only one size angle that meets at every vertex, the vertex angle of the regular polygon must divide 360°. This is true of all three of the above polygons. However, the vertex angle of a pentagon is $180 - \frac{360}{5} = 180° - 72° = 108°$. Since 108° does not divide 360° evenly, it cannot tessellate the plane by itself. Second, the vertex angle cannot be greater than 120°. This is because at least three polygons must meet at a vertex to make a tessellation (think about what would happen if only two regular polygons met at a vertex). The vertex angles of regular polygons become larger as the number of sides increases. Try this out yourself by sketching some regular polygons or by using toothpicks to make regular polygons with increasing numbers of sides. You can also look at the sequence of vertex angles as the number of sides increases:

Number of sides of regular polygon	3	4	5	6	7	8	...
Degree of vertex angle of regular polygon	60	90	108	120	128.57	135	...

Notice that the size of the vertex angle increases in the table as the number of sides of the regular polygon increases. Although this is not a proof that the angles continue to increase, it is certainly evidence of it.

The three pictures below represent the only three regular tessellations:

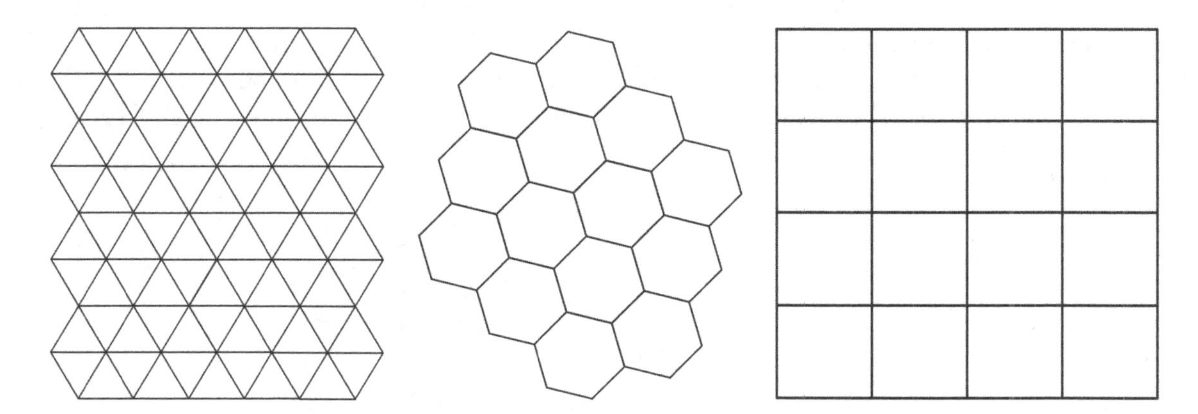

A tessellation is called a semi-regular tessellation if there is more than one type of prototile, all of the prototiles are convex regular polygons, and all of the vertex figures are congruent to each other.

There are eight semi-regular tessellations of the plane, all of which are shown in the figure below:

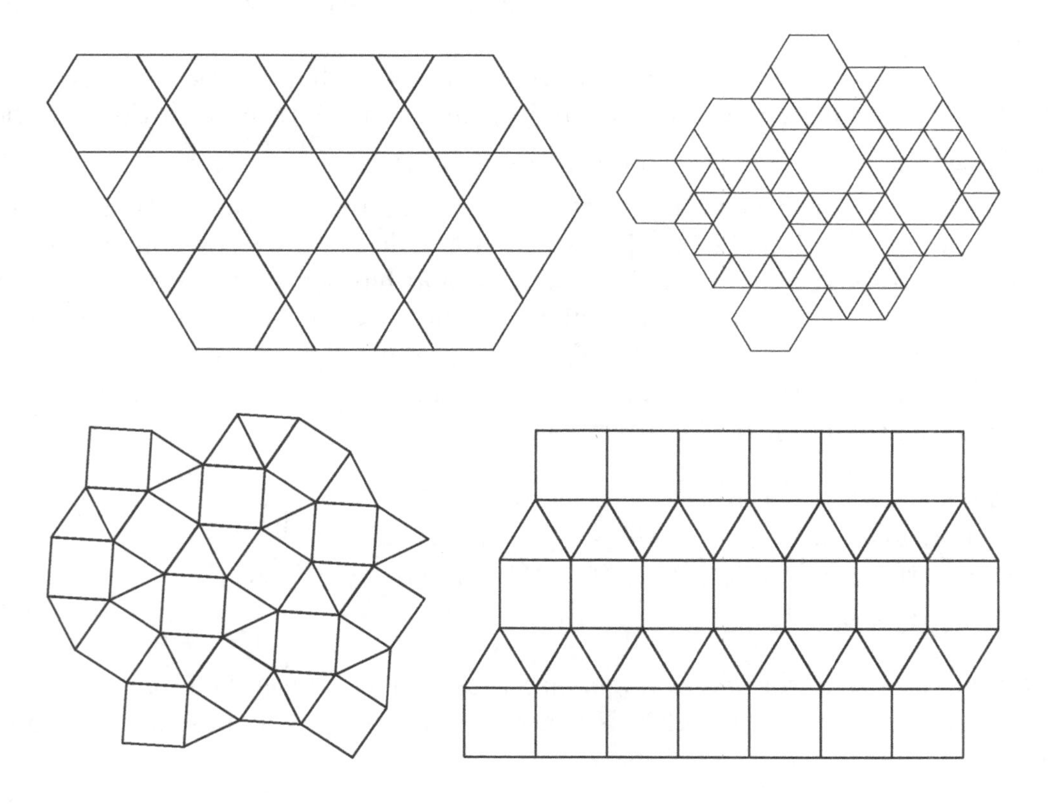

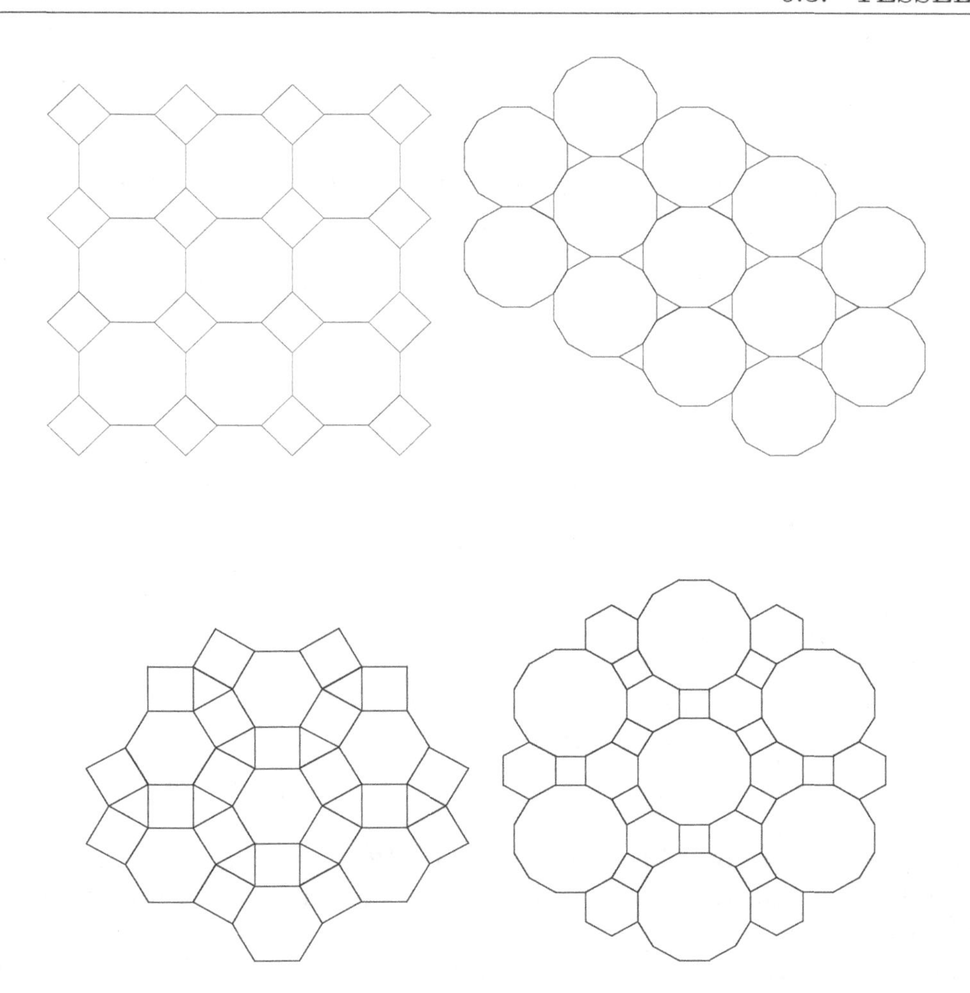

The reason that these are the only semi-regular tessellations of the plane is somewhat more complicated, and we will not discuss it here.

A dual tessellation is a tessellation formed by putting vertices in the center of each polygon of a tessellation and drawing edges between the centers of adjacent polygons.

For instance, in the series of pictures on the next page, the first picture in the first row is a tessellation (a semi-regular tessellation of octagons and squares), the second picture in the first row is the tessellation with all of the centers of the polygons shown. The first picture in the second row shows the dual tessellation of the original tessellation in red, and the second picture in the second row is the dual tessellation without the original tessellation shown. Note that the dual of a semi-regular tessellation is not necessarily a semi-regular tessellation; the dual tessellation in this case is monohedral (made up of only one prototile), and the prototile is also not a regular polygon.

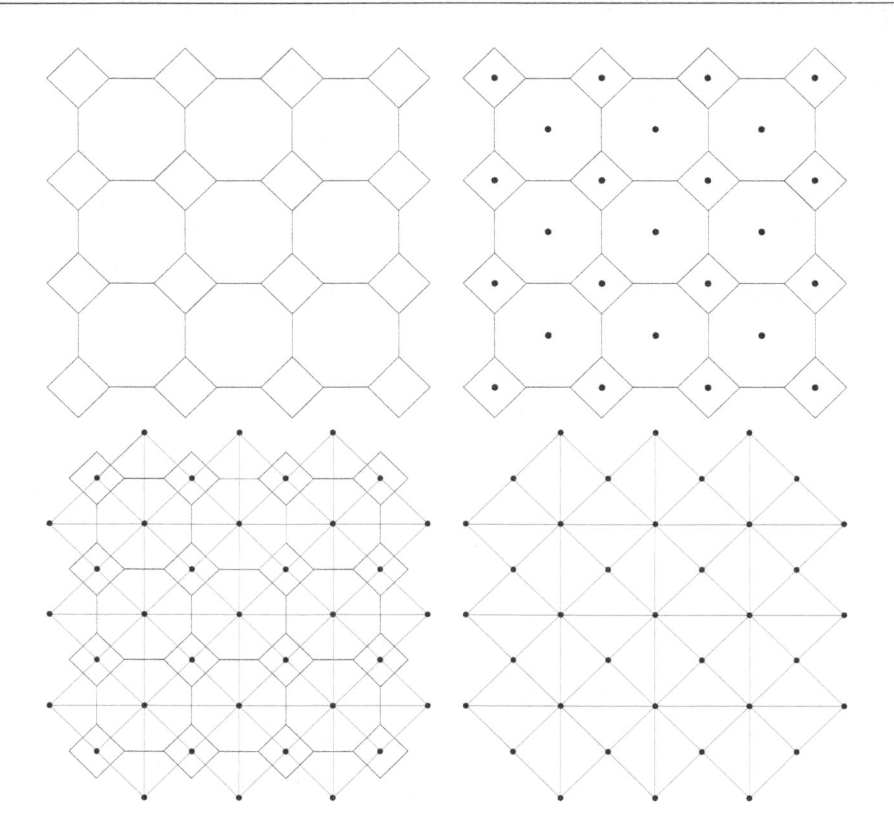

Not all tessellations are regular or semi-regular. Both regular and semi-regular tessellations have regular polygons as prototiles; any tessellation made up of prototiles that are not regular polygons is neither regular nor semi-regular.

Any quadrilateral can produce a monohedral tessellation. Why? Recall that if you add the degrees of all of the vertex angles in a quadrilateral, you get 360°. This was discussed at the end of Section 4.2. Also, a complete circle is 360°. Therefore, if you take four copies of the same quadrilateral and arrange them so that a different angle of the quadrilateral meets at a vertex, the angles will fit together around the vertex without any overlapping or gaps. If we can do that for each vertex of a tessellation, then we have found a way to tessellate the plane with a particular quadrilateral.

For example, start with a quadrilateral, such as the quadrilateral pictured below.

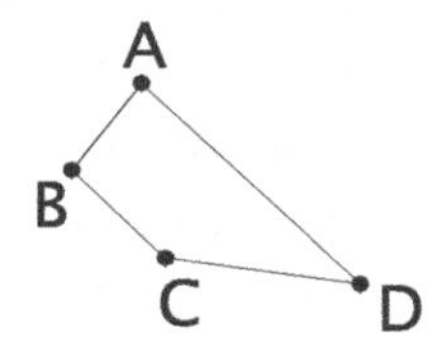

Next, find a midpoint of an edges of the quadrilateral, as shown below.

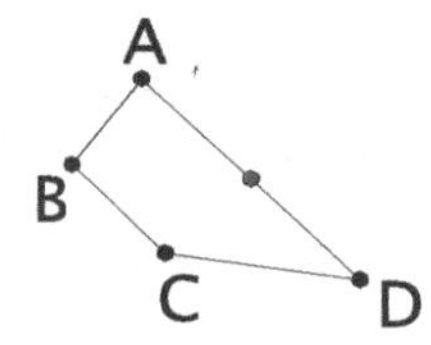

Then, rotate the quadrilateral around the midpoint of the side $\overline{AD}$ 180°, as pictured below. When a figure is rotated around a point, label the new vertices the same as the original figure, but with an additional $'$ after the name. Therefore, Point A, when rotated around a point, becomes Point A'.

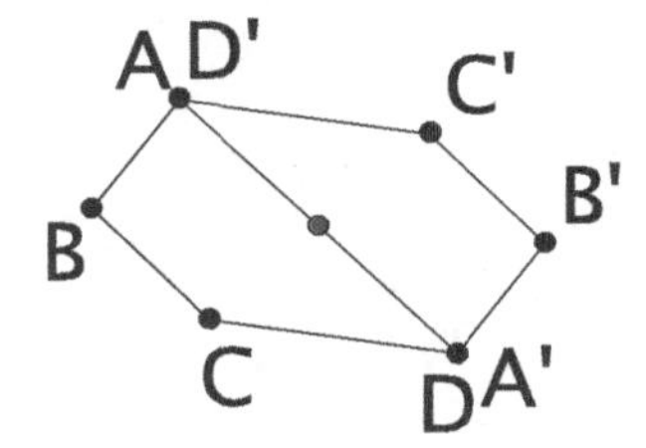

Next, find the midpoint of side $\overline{A'B'}$ of the rotated figure and rotate the new quadrilateral 180° around side $\overline{A'B'}$, as shown below. Now, Point A' becomes A'' when rotated.

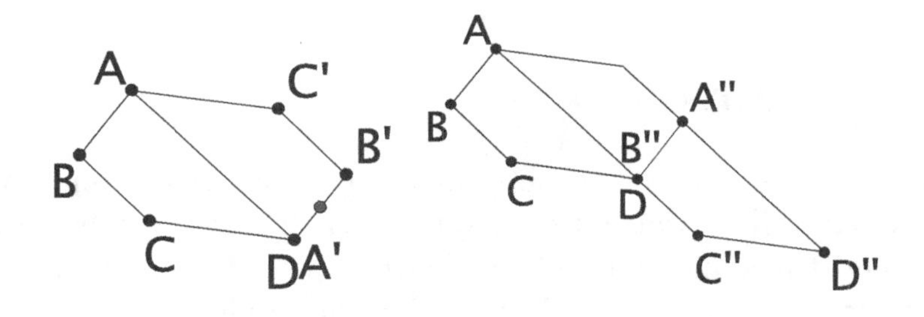

Finally, find the midpoint of side $\overline{B''C''}$ of new quadrilateral and rotate the new figure 180° around side $\overline{B''C''}$, as shown below:

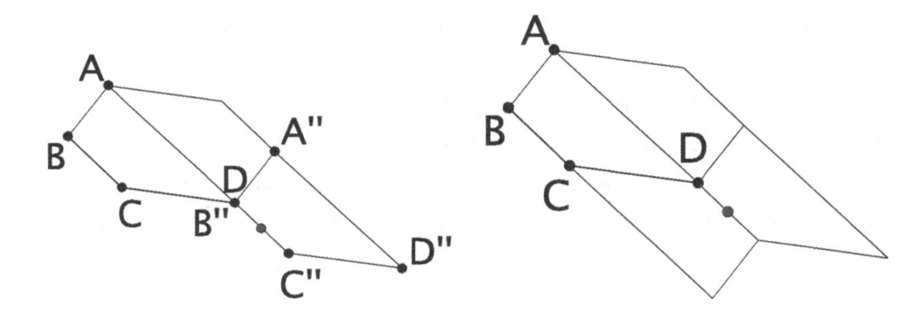

Now all of the space around Vertex D has been filled. Use the same procedure with the midpoints of other edges to create the rest of the tessellation. The result is pictured below.

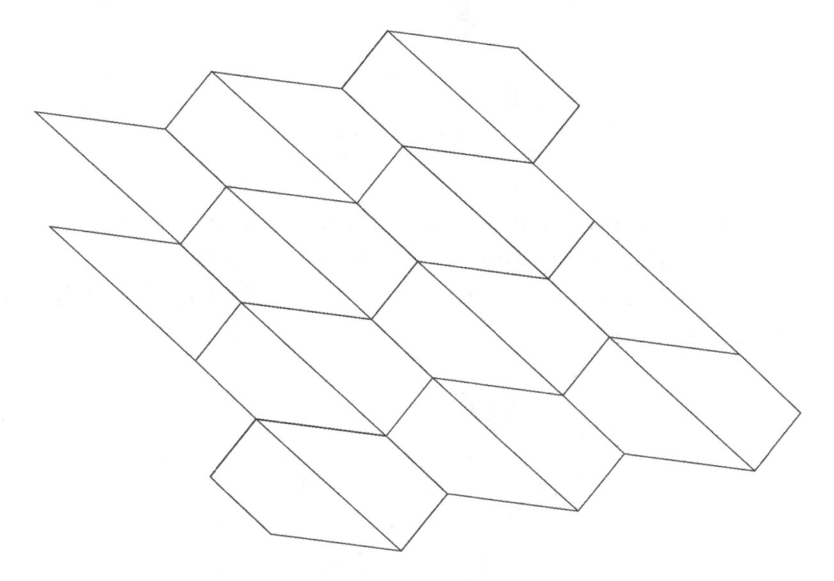

This procedure will work for any quadrilateral. Notice that for each vertex of the tessellation above, each angle of the quadrilateral appears once.

In the next section, we will discuss isometries, which will lead us into symmetries of tessellations.

6.4 Isometry Movements

Tessellations have the same symmetries as regular polygons (rotational symmetry and reflectional symmetry) as well as translational symmetry. These motions will be explained in more detail below, and the symmetries of tessellations will be demonstrated in Section 6.5. The motions of rotation, reflection, translation, and glide-reflection are called isometries because they preserve the distance between points. This means that if you take a square and rotate it, reflect it, or translate it, it will remain a square and stay the same size. In fact, if you take a picture containing both square and octagons and rotate, reflect, or translate the whole thing, the squares and octagons will not only stay the same size, but they will also be the same distance from each other.

As discussed in Section 6.1, it is necessary to show both the center of rotation and the degrees of rotation for any rotation. It is now more important than ever to specify the center of rotation because an object can be rotated around a point that is not the center of the object itself. The center of rotation is not the same thing as the center of a particular object.

For instance, we will rotate the object below 180° around the point O.

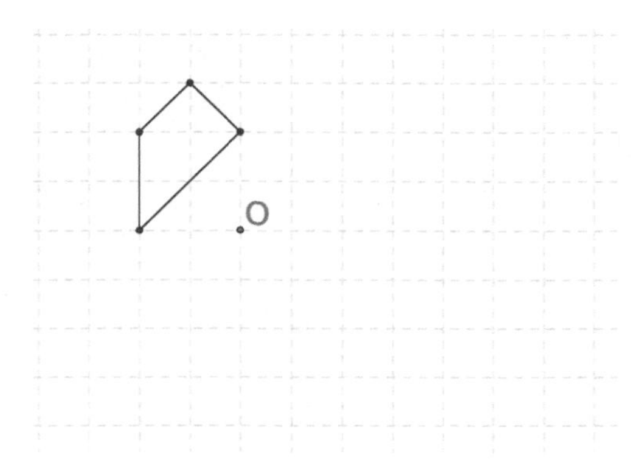

To do this, we will rotate each vertex of the object around the point O by 180°. Recall that a 180° turn is a straight line (Section 3.1). Therefore, we want to rotate each vertex (call the given vertex V) of the object around point O so that it lands on the straight line $\overleftrightarrow{VO}$, but on the other side of O the same distance from O that V is from O.

To illustrate, consider point A in the figure below. Draw a straight line (either on paper or in your mind) through A and O. The point O is to the right 1 unit, down 3 units from the point A, so our new point should be to the right 1 unit, down 3 units from the point O. The rotated point is labeled as A' in the figure below.

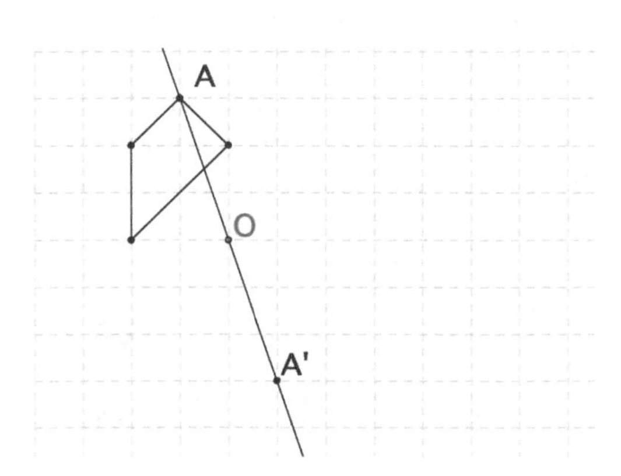

Now do the same for the other three vertices of the object, and then connect the vertices. The figure below shows all of these steps.

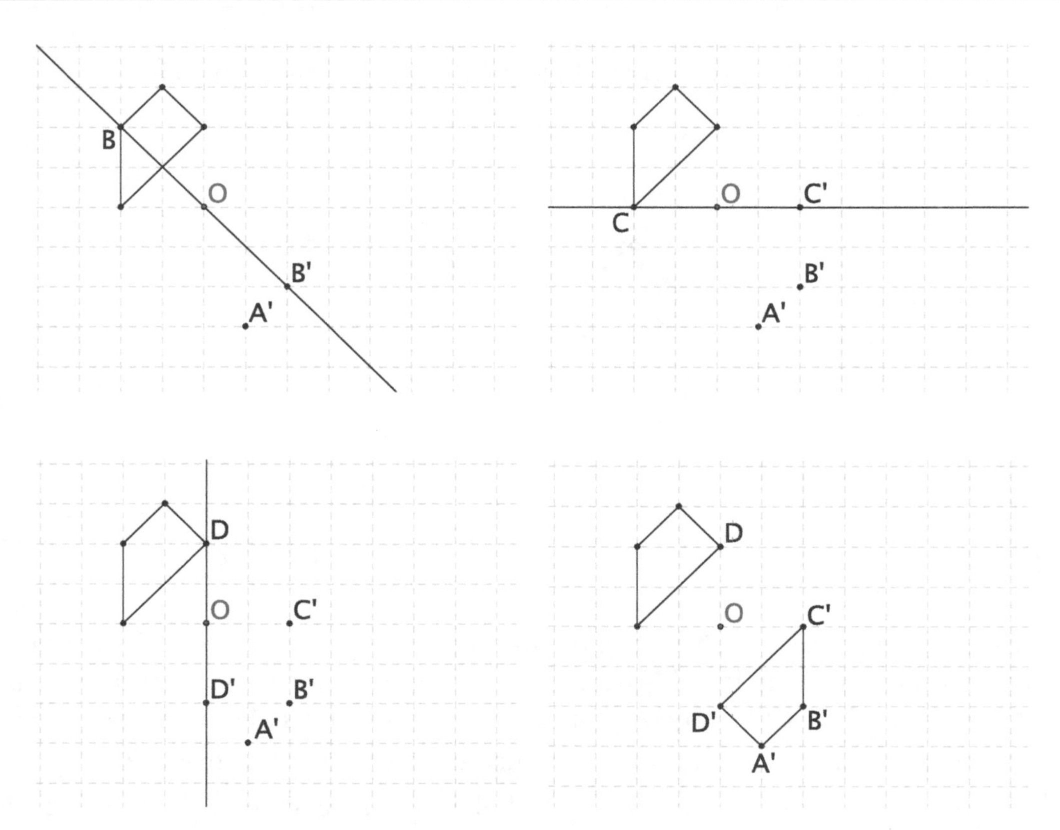

Doing rotations by hand can be done with any angle. For instance, say we wanted to rotate that same object by 90° counter-clockwise around the point O. Then, for each vertex (or important point) of our figure we would draw a line segment from the vertex (that point) to O. Then, we would draw another line 90° counter-clockwise from that line segment. The rotated point would be on this line an equal distance from O as the original vertex (point) was.

To illustrate, we again begin with point A and draw a line segment from A to O (first picture below). Then, we draw a ray 90° from line segment $\overline{AO}$ with O as the vertex of the angle. The rotated point A' is on that ray the same distance from O as A is from O.

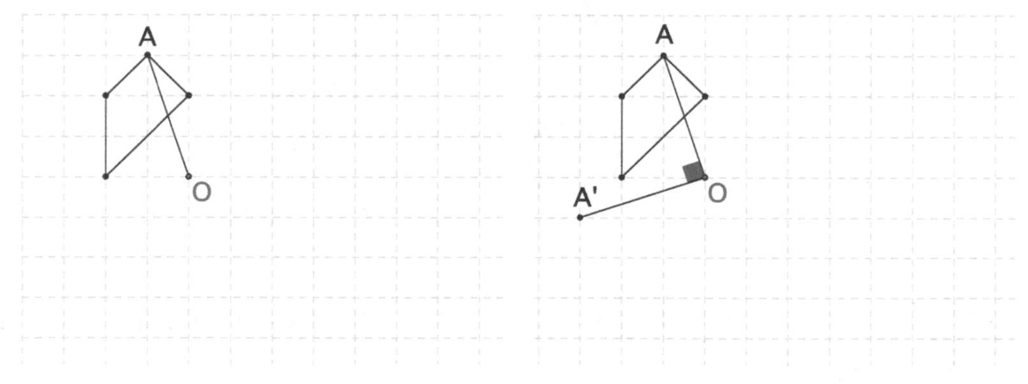

Continue this process for all of the vertices of the object; then connect the vertices to recreate the object in its new location. This is illustrated below.

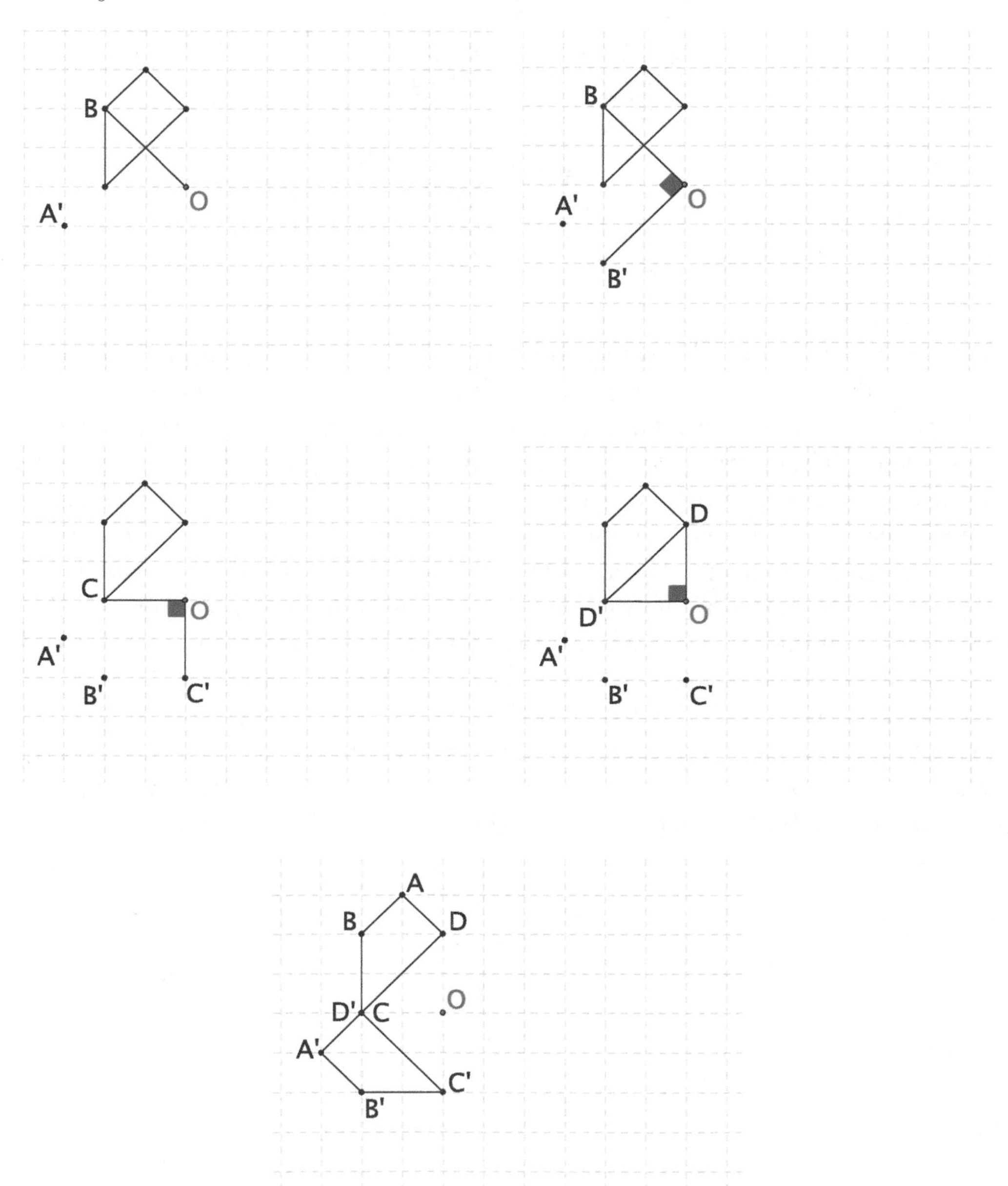

Reflection occurs when an object or the entire plane is moved across a line (called the reflection line) in such a way that each point in the reflected object is the same distance from the reflection line as the original point was, and any line containing both an original point and its reflected point is perpendicular to the reflection line.

This is illustrated in the picture below.

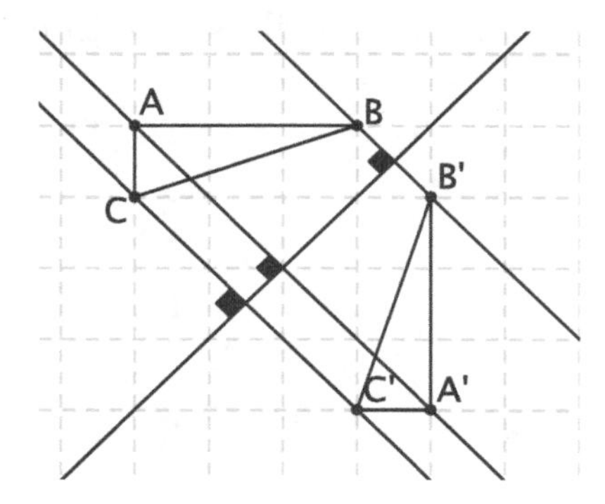

To reflect an object across a line of symmetry, reflect each of the important vertices of the object across the line of symmetry, and then connect the points to re-construct the object. To reflect the points, begin by drawing a line through the point in question perpendicular to the line of reflection. Then, draw a point on that line on the opposite side of the reflection line which is equal in distance from the reflection line as the original point is from the reflection line. Continue in this way for each point and then connect the points to recreate the object.

Translation occurs when an object or the entire plane is moved across the plane such that the size and orientation of the object is the same; only the position has changed. To show translational symmetry, we need to draw a vector that represents that movement. In this case, the position of the object or objects on a plane matters.

A vector is an arrow that indicates both length and direction. The vector in the figure below represents a distance equal to the length of the arrow, and a direction equal to the direction the arrow is pointing. Seeing this vector, I know that this vector represents a movement from the original position to the right 3 units and down 2 units.

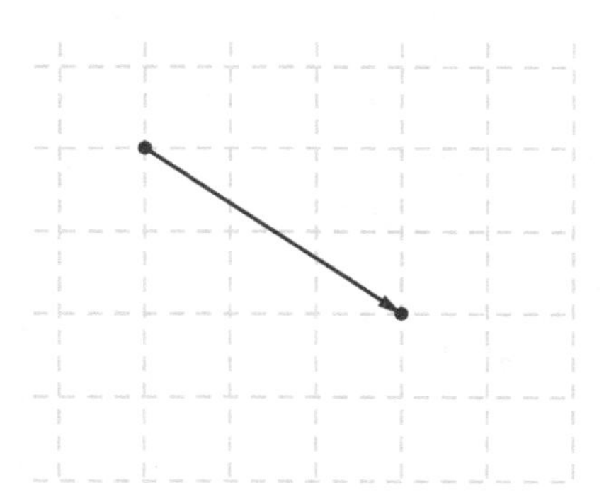

Notice that the vector has an arrow at one end. The tip of the arrow must be at the point where the vector ends.

In the figure below, the first picture shows an object in the plane (the object is red). The second figure shows a vector specifying a translation of the object. The third picture shows the original object (in red) and the resulting translated object (in purple). To do a translation by hand, translate each important point of the object, like its vertices, and then connect the points to recreate the figure.

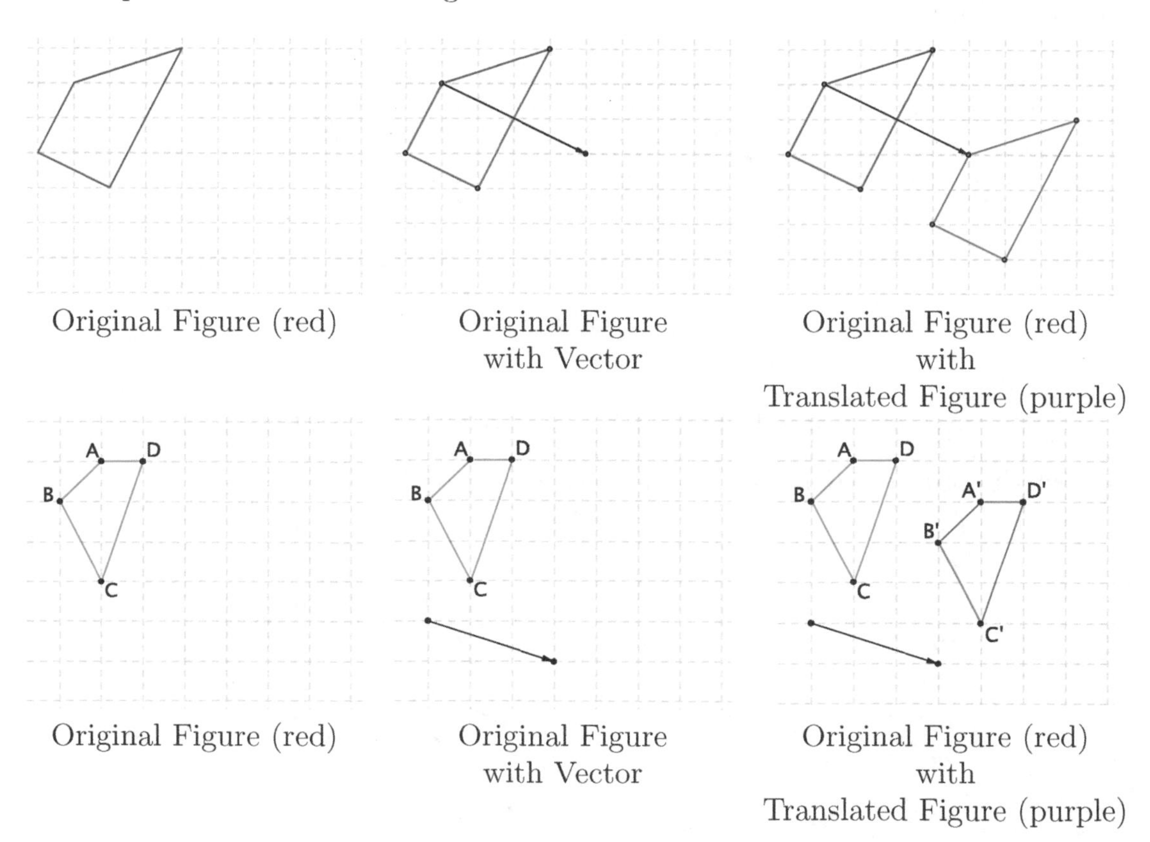

Original Figure (red) | Original Figure with Vector | Original Figure (red) with Translated Figure (purple)

Original Figure (red) | Original Figure with Vector | Original Figure (red) with Translated Figure (purple)

Understanding the Text:

1. Rotate the objects below the given number of degrees around the given center of rotation (O).

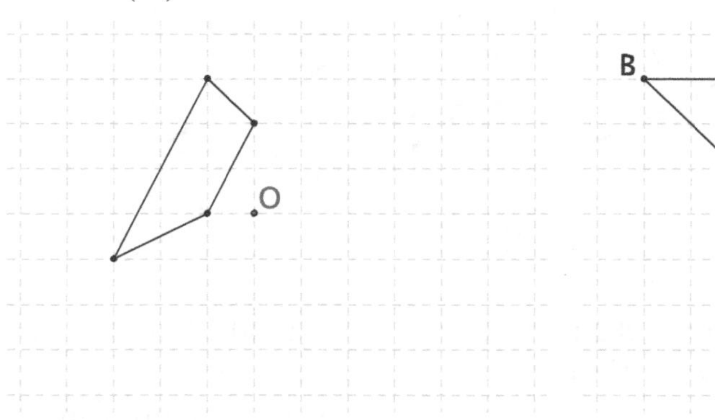

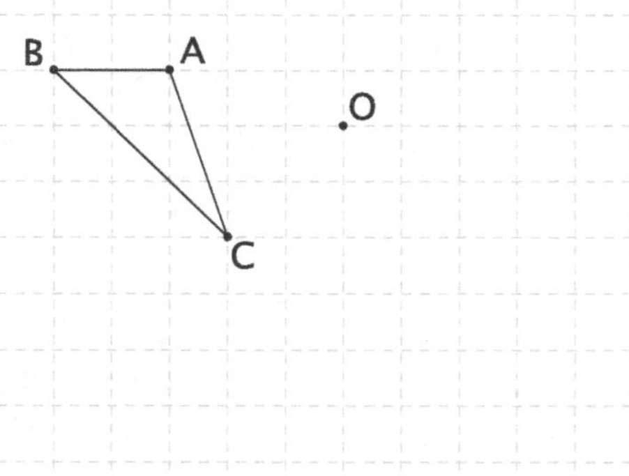

Rotate 180° around O Rotate 90° around O

2. Reflect the objects below across the given reflection line.

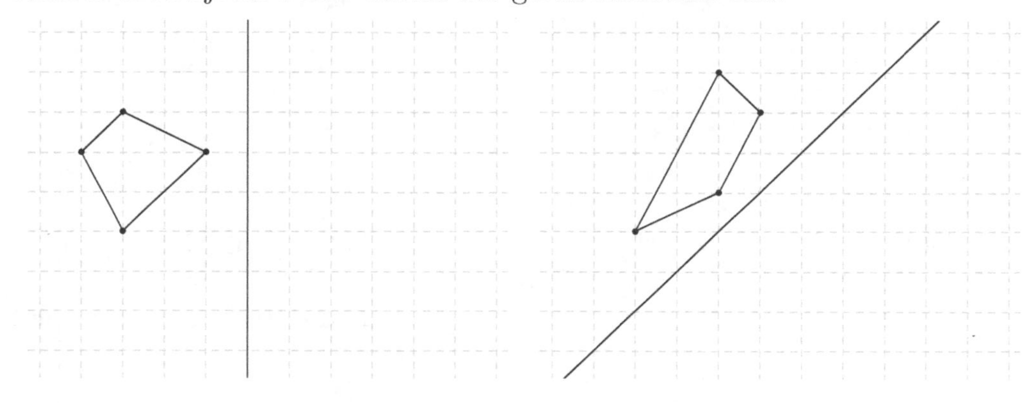

3. Translate the figure below by the given vector.

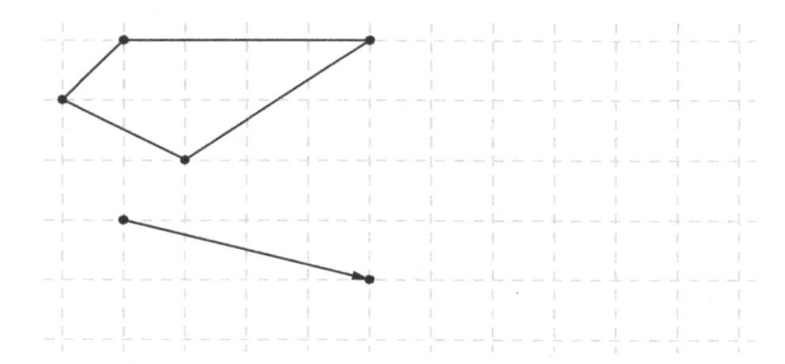

Some notes on translating objects:

(a) Translate the vertices of the object you want to translate. Then, use a straight edge to connect the points.

(b) The placement of the vector does not matter. A vector simply gives you the instructions for how far and in what direction to move the object. In this case, each point on the object should move to the right 4 units and down 1 unit.

4. Translate the figure below by the given vector.

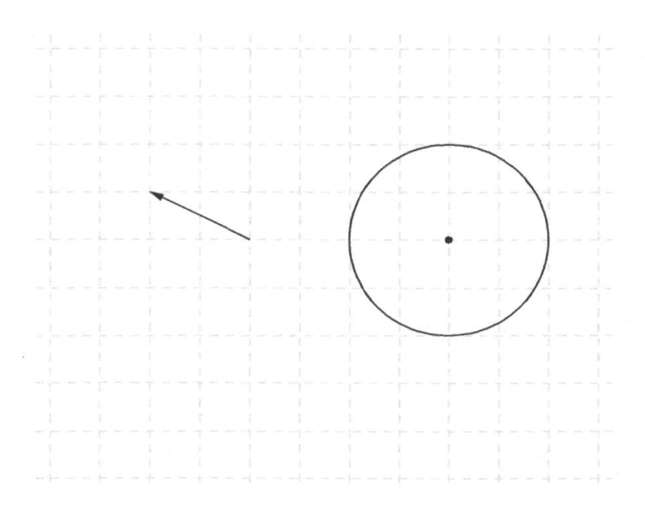

Some notes on translating objects:

(a) Translate the important points of the object you want to translate. Then, draw the object using those points. In this case, you could either translate the center and a point on the circle and draw the circle with a compass, or you could translate the four points on the circle that lie on the grid and sketch a circle using those points.

(b) The placement of the vector does not matter. A vector simply gives you the instructions for how far and in what direction to move the object. In this case, each point on the object should move to the left 2 units and up 1 unit.

5. Translate and then reflect the triangle below by the given vector and reflection line.

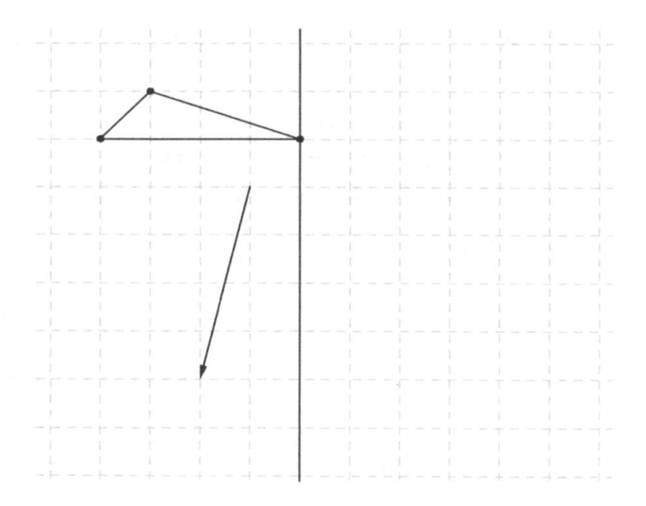

Some notes on working through this problem:

(a) Translate the important points of the object you want to translate and then create the object from those points.

(b) The placement of the vector does not matter. A vector simply gives you the instructions for how far and in what direction to move the object.

(c) Once you have translated the object, you should reflect the object over the line. In this case, you are given a vertical line, so all you have to do is count horizontally from each important point of the translated object to the line. Then, count horizontally the same number of units on the other side of the line. That is where the reflected object should be.

(d) Connect the important points you reflected in the previous step to recreate the object.

6.5 Composition of Isometry Movements

As in Section 6.2, we can consider the isometry movements as elements of a group and "followed by" as an operation of the group. An isometry movement is a motion that makes the prototiles of a tessellation land on prototiles of the same size and shape. Any isometry movement of a tessellation can be described by a reflection, rotation, translation, or some combination of these motions.

What happens if we do a reflection followed by another reflection?

The first picture below shows an object along with two reflection lines that are parallel to each other; one of the lines is labeled r and the other is labeled s. We will first reflect the object over r and then over s. The second picture shows the object reflected over r, and the third picture shows the result of reflecting the reflected object over s. Each point of the reflected object is labeled with a "prime" or $'$ symbol next to each vertex. For instance, Point A becomes Point A'. When A' is reflected over line s, the point becomes A''.

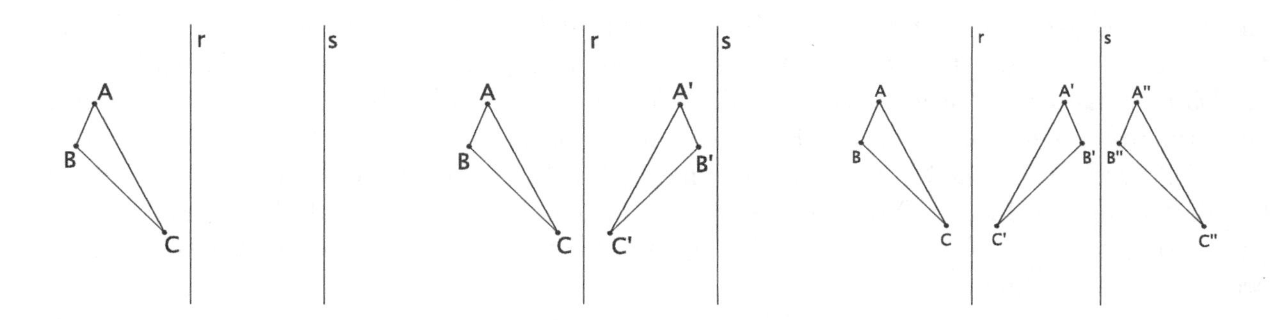

So what happened? The resulting object, in this case $\Delta A''B''C''$, is the same size and in the same orientation as the original object, but it is in a different position. Therefore, it appears that when the two reflection lines are parallel to each other, a reflection followed by another reflection is the same as a translation in a direction perpendicular to the reflection lines. Let us look at another example, this time on a grid, so we can also try to figure out the distance of the translation.

Note that in the case below, $A'B'C'D'$ overlaps s. This is not a problem; we just reflect the points that are on the right side of s over to the left side of s and the points that are on the left side of s over to the right side of s. This is illustrated in the first picture on the following page.

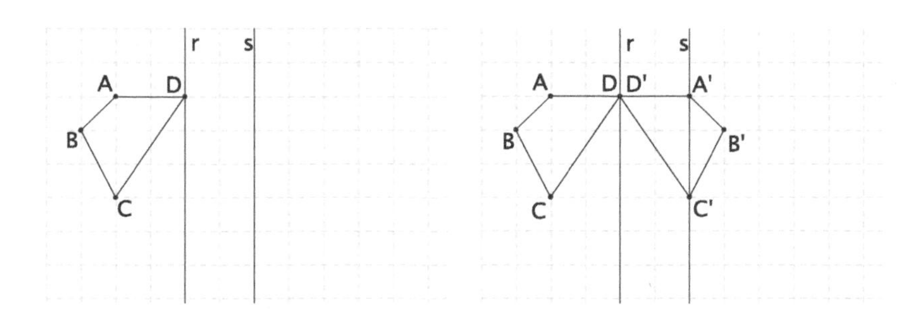

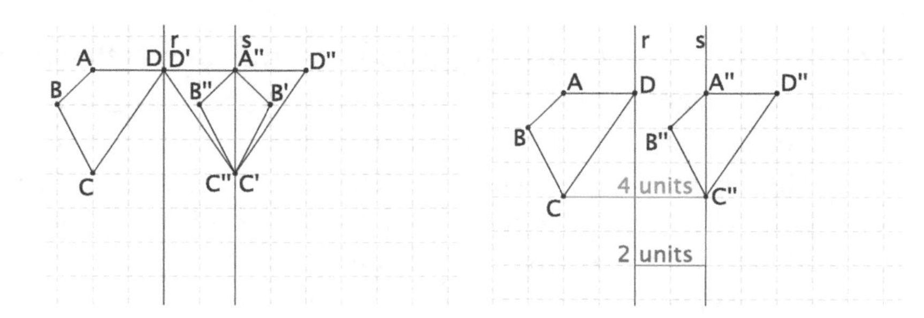

As you can see, if the lines of reflection are 2 units apart, then the original object and the resting object are 4 units apart. It appears that the distance of the translation is twice the distance between the two reflection lines. Let us explore this to see if it is true.

First, consider an object, say ΔABC. We know that any object and its reflection are the same distance from the reflection line. Therefore, the distance between ΔABC and the first reflected object $\Delta A'B'C'$ will be twice the distance between the original object and the first reflection line, r. Now the distance between the first reflected object $\Delta A'B'C'$ and the second reflected object $\Delta A''B''C''$ will be twice the distance between $\Delta A'B'C'$ and the second reflection line, s.

Also note that a reflected object is simply the reflection of all the points on that object. In particular, we can reflect the important points (such as the vertices) of the object and then connect the reflected vertices to recreate the object. Therefore, all we need to show is that points reflected over one line and then another line parallel to the original line behave the way we expect them to.

We first consider the case where the first reflected point (say X') is between the original point (X) and the resulting point (X''); then the original point (X) and the resulting point (X'') are on opposite sides of both r and s. The sum of the distance between ΔABC and $\Delta A'B'C'$ and the distance between $\Delta A'B'C'$ and $\Delta A''B''C''$ is the distance between ΔABC and $\Delta A''B''C''$. The distance between ΔABC and r is the same as the distance between r and $\Delta A'B'C'$. Therefore, the sum of the distance between ΔABC and r and the distance between $\Delta A'B'C'$ and s is equal to the distance between r and s. Therefore, the distance between ΔABC and $\Delta A''B''C''$ is twice the distance between r and s.

The pictures below illustrate the previous paragraph.

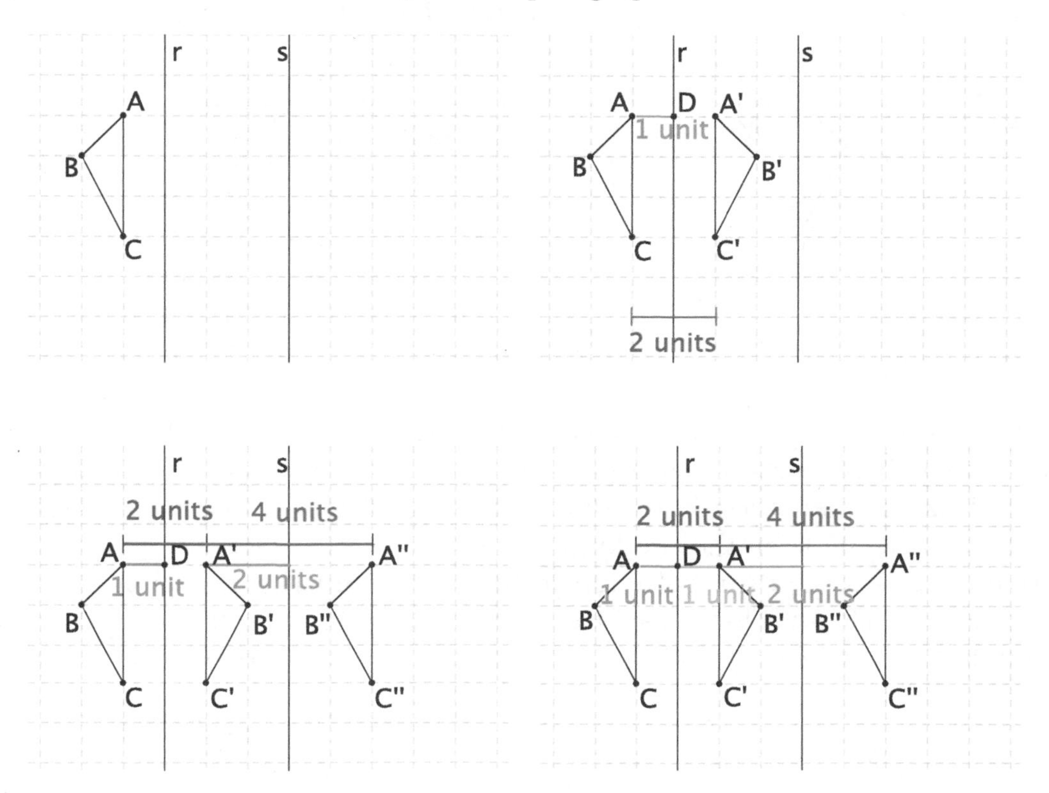

The second case is when the first reflected point (X') does not lie between the original point (X) and the resulting point (X''). In this case, the distance between the original point (X) and the resulting point (X'') is equal to the **difference** between the distance from the original point (X) to the first reflected point (X') and the distance from the first reflected point (X') to the second reflected point (X''). Also in this case, the distance between r and s is equal to the difference between the distance from the original point (X) to the first reflection line s and the distance from the first reflected point (X') and second reflection line (s). As in the previous case, each of the first set of distances is twice the second set of distances. Since in general

$$\frac{a}{2} - \frac{b}{2} = \frac{1}{2}(a - b)$$

this means that the distance between X and X'' is twice the distance between r and s.

We illustrate this case below using the point B; this figure is one of the examples given earlier in this section. Note that B' is not between B and B''.

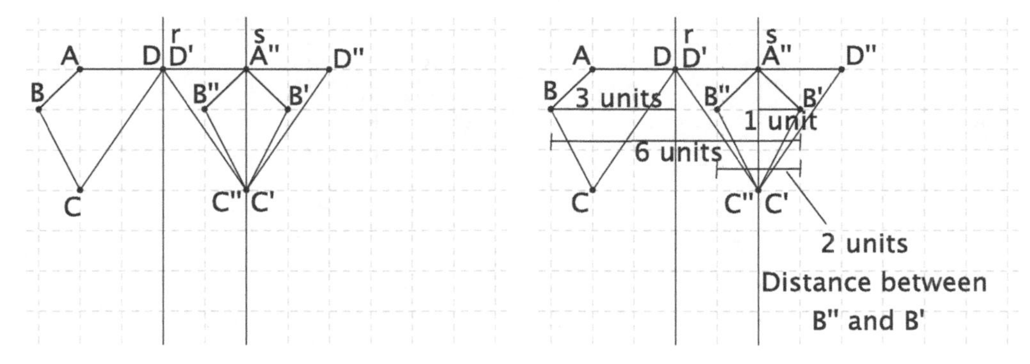

Since these are the only two possible cases, we have shown that a reflection of an object over a line followed by a reflection over a line parallel to the first line is the same as a translation of the original object in the direction perpendicular to the lines a distance equal to twice the distance between the two lines.

What happens if we do a rotation followed by another rotation?

Now let us consider what happens if the reflection lines r and s intersect and we want to reflect an object over r followed by a reflection over s. An example is given below:

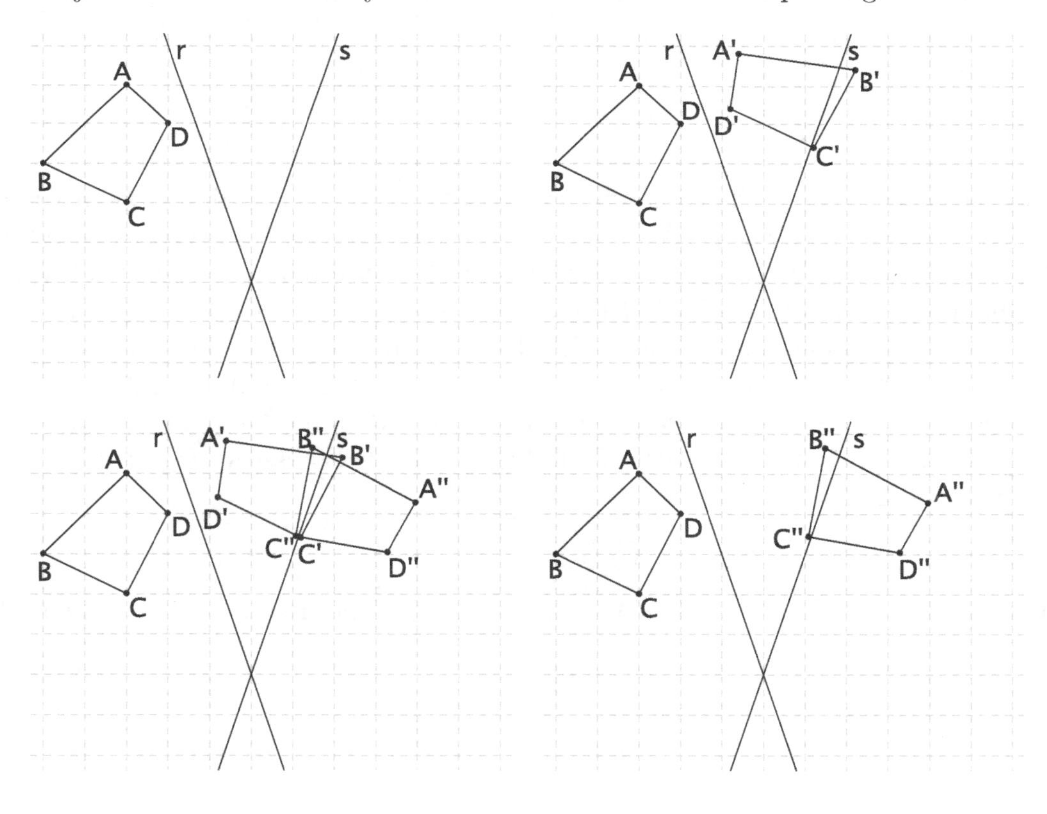

What one isometry would take you directly Quad $ABCD$ to Quad $A''B''C''D''$? We can see that the orientation is not the same, so it cannot be a translation. Also, the figure does not appear to be "flipped," that is, A and D are on the same side of Quad $ABCD$ as A'' and D'' are of Quad $A''B''C''D''$. Is this a rotation? The shape certainly appears turned.

If it is a rotation, we now need to find the center of rotation and the degrees of rotation. The center of rotation would be a point where each point (say X) is equidistant from the center of rotation (call it O) as O is from X''. The point of intersection between r and s appears to be such a point. Let us check the distances for this example:

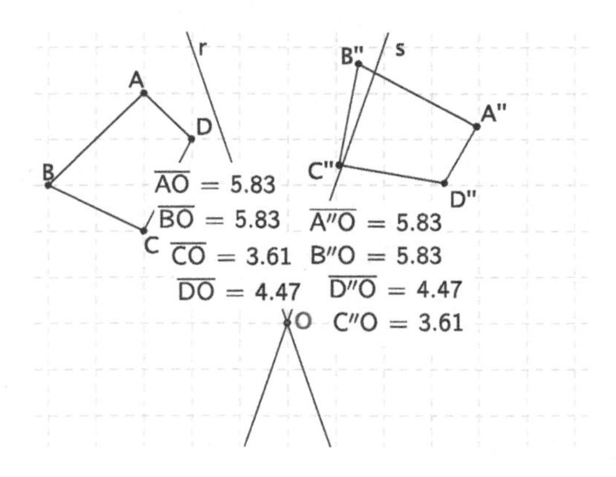

So each of the vertices are the same distance from O as their reflections are from O.

Now let us see how far the quadrilateral is rotated. To do this, draw a ray from O through D and from O through D''. Then measure angle DOD''. This is the angle of rotation of the quadrilateral, which is shown in red in the picture below. You can do the same with A and A'' or any other pair of original and resulting points; you will get the same measurement.

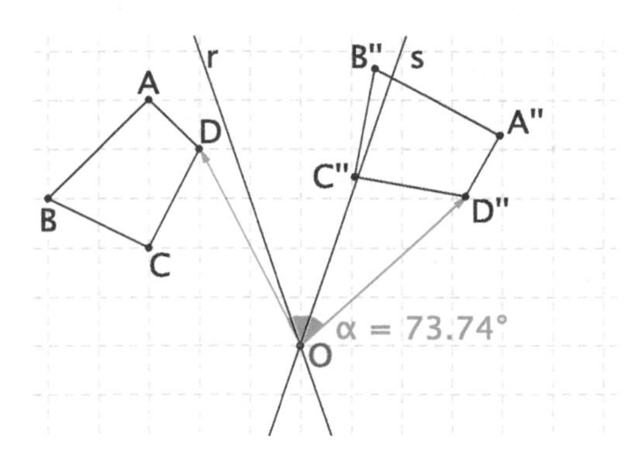

Now let us compare the angle at which the quadrilateral was rotated to the angle between r and s. This angle is in red in the picture below.

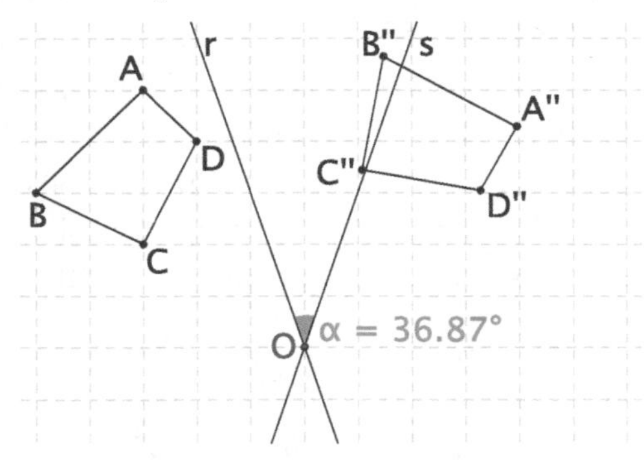

As you can see, the angle at which the quadrilateral was turned is twice the angle between r and s. The reason for this is similar to the reason for why if r and s are parallel to each other: the figure is translated twice the distance between r and s. We will not go into detail here, however.

Understanding the Text:

1. Reflect the triangle ΔABC below over line r and label the triangle you drew $\Delta A'B'C'$. Now reflect $\Delta A'B'C'$ over line s and label that triangle $\Delta A''B''C''$. What one isometry is equivalent to these two motions. That is, what one isometry movement would you use to go directly from ΔABC to $\Delta A'B'C'$? Describe the isometry in detail. For instance, if it is a translation, give the distance and direction; if it is a rotation, give the center of rotation and the angle of rotation.

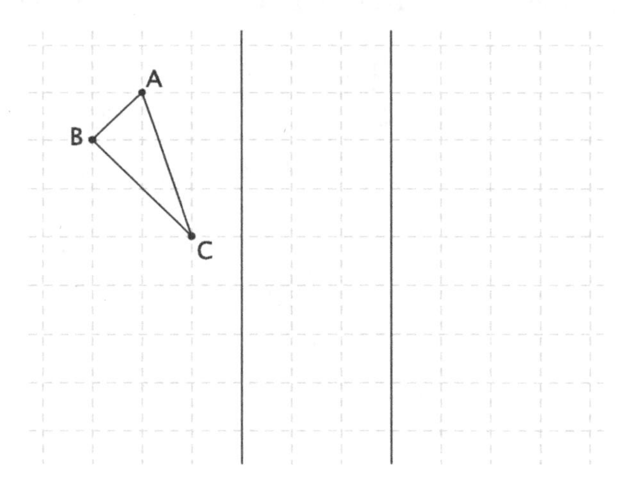

2. Reflect the triangle ΔABC below over line r and label the triangle you drew $\Delta A'B'C'$. Now reflect $\Delta A'B'C'$ over line s and label that triangle $\Delta A''B''C''$. What one isometry

is equivalent to these two motions. That is, what one isometry movement would you use to go directly from ΔABC to $\Delta A''B''C''$? Describe the isometry in detail—if it is a translation, give the distance and direction; if it is a rotation, give the center of rotation and the angle of rotation; if it is a reflection, give the reflection line; if it is a glide-reflection, give the distance and direction of the glide and the reflection line.

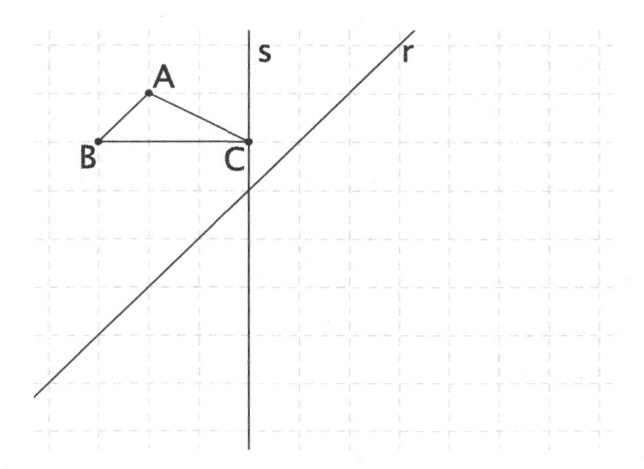

3. Consider the figure below and answer the questions that follows. In the figure, Point K is the center of the pentagon; Points J, F, G, H, I, and J are the midpoints of line segments $\overline{ED}$, $\overline{DC}$, $\overline{CB}$, $\overline{BA}$, and $\overline{AE}$, respectively. All rotations are in the counterclockwise direction.

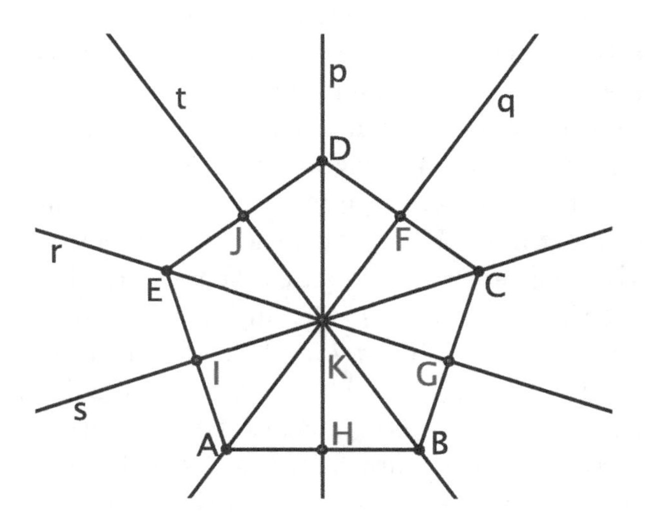

   (a) If you reflect Point J over line s, Point J will land in the position originally occupied by what point?

(b) If you reflect Point A over line q, Point A will land in the position originally occupied by what point?

(c) If you rotate Point H around Point K an angle of 288°, Point H will land in the position originally occupied by what point?

(d) If you rotate Point E around the center of the pentagon an angle of 360°, Point E will land in the position originally occupied by what point?

(e) If you reflect Point J over line p followed by line t, Point J will land in the position originally occupied by what point?

(f) If you reflect Point E over line p followed by line t, Point E will land in the position originally occupied by what point?

(g) If you rotate Point E around Point K an angle of 144° followed by a rotation of 216°, Point E will land in the position originally occupied by what point?

(h) If you rotate Point I around Point K an angle of 72° followed by a reflection over line s, Point I will land in the position originally occupied by what point?

6.6 Symmetry of Tessellations (Isometry Symmetries)

We have already discussed rotational and reflectional symmetry of objects in Section 6.1. Now, however, we will discuss the symmetry of the entire plane. For example, we will look at the symmetries of tessellations, which are explained in Section 6.3. The symmetries of tessellations are isometries, which were discussed in Section 6.3; an isometry is a symmetry of the tessellation if the movement leaves the tessellation looking exactly as it did originally—in the same position, in the same orientation, and the same size as before.

One example of a symmetry of a tessellation is translational symmetry of a tessellation. Translational symmetry of a tessellation occurs when a translation movement leaves the tessellation looking exactly as it did originally. That is, the translation movement puts shapes of exactly the same size and shape in each position of the tessellation.

For example, consider the tessellation on the far left below. We want to translate the figure so that all of the shapes land on identical shapes. In this case, all of the triangles in the figure are the same size and shape, so we only have to worry about triangles landing on triangles. If we start with a vertex of a triangle, for instance, we need to find an analogous point on another triangle to send it to. The second figure below shows a chosen point; the point is at the top of a triangle. To find translational symmetry, we need to move that point to the top of another triangle. The third figure shows such a vector, which takes point A to an analogous point. This point has triangles on all sides, in the same configuration as in the original picture. If the entire tessellation is moved by the same vector as the one between Point A and Point B, the tessellation will look exactly the same.

This is illustrated in the series of pictures below.

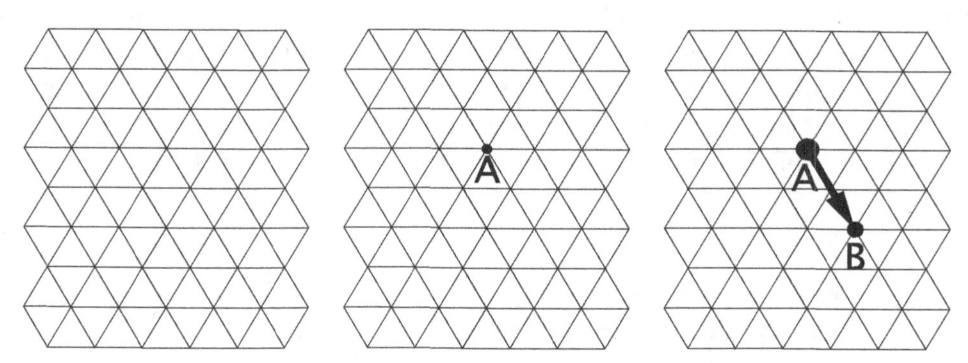

Reflectional symmetry of a tessellation occurs when there is a line of symmetry that will divide the entire tessellation in half so that the tessellation on one side of the line is the mirror image of the tessellation on the other side of the line. Typically, we only give one representative line of reflection for each type of reflectional symmetry, since if there is one line of reflection of a tessellation, there are usually infinitely many lines of reflection. This is because most tessellations are repetitive. The figure below gives a tessellation and then all of the representative lines of symmetry of that tessellation. Remember that tessellations are infinite, so the picture shows only a portion of the entire tessellation. This is why some of the lines of reflection are not lines of reflection of the picture shown; they are lines of reflection of the entire tessellation.

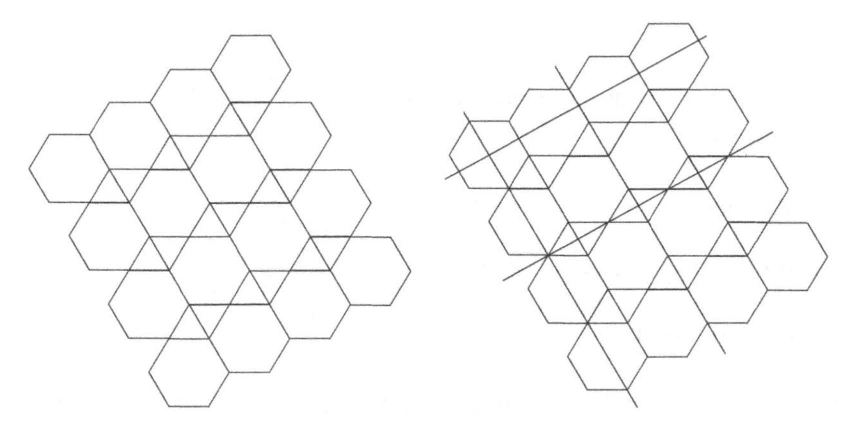

Another type of symmetry of a tessellation is rotational symmetry. Rotational symmetry of a tessellation occurs when rotation of a particular number of degrees around a particular point make the tessellation look like it did originally. To show the rotational symmetry of a tessellation, we must show all representative centers of rotation where if the tessellation is turned on that center there are positions in which the tessellation looks the same; then we must indicate all of the possible degrees for which the tessellation looks the same for each of the centers of rotation.

For example, consider the tessellation below. The first picture shows the tessellation. The second picture shows the tessellation with all of the centers of rotation; only representative centers of rotation are shown because if there is one center of rotation in a tessellation, there are usually infinitely many copies of that center of rotation. The centers of rotation are shown in red. The third picture shows the tessellation with the centers of rotation and the number of degrees of rotational symmetry for each center of rotation.

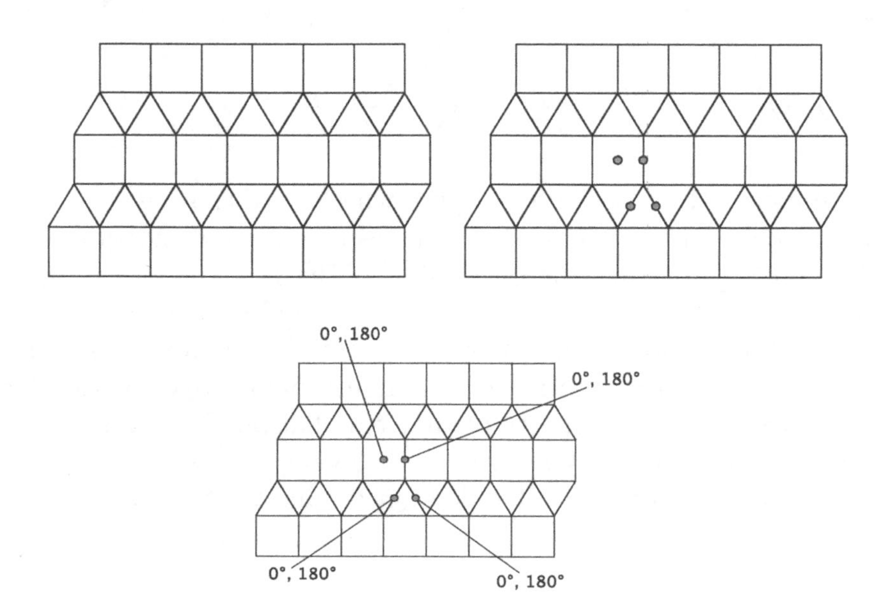

When looking for centers of rotational symmetry of a tessellation, there are three places you should look for them:

1. Centers of prototiles of the tessellation.

2. Midpoints of edges of the tessellation.

3. Vertices of the tessellation.

As in the example above, not all tessellations will have centers of rotational symmetry in all three of these places. Also, just because a prototile has a particular center of rotational symmetry does not mean the entire tessellation has that center of rotational symmetry. All shapes must land on shapes of their own size and type. In the example above, we see that if the tessellation were turned 90° around the center of a square, triangles would land on squares and the tessellation would be changed. That is why the center of the square in the above tessellation has only 0° and 180° symmetry, even though the square by itself has more angles of rotational symmetry.

Understanding the Text:

1. Show one translational symmetry of the tessellation below. You may use any starting position of your choice.

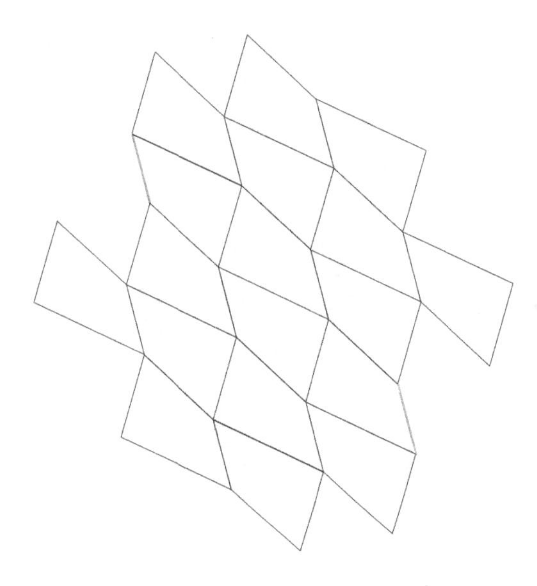

2. Show the different lines of symmetry of the tessellation below. Two lines of reflectional symmetry are the same if they have the same slope (go in the same direction) and go through the same objects of the tessellation. If this is not the case, then the lines are different.

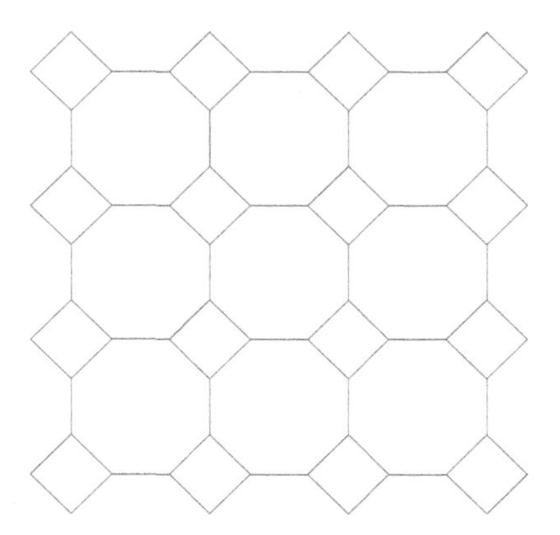

3. Indicate the rotational symmetry of the tessellation below. Show the rotational symmetry for one example of each type of center. For instance, if you give the rotational symmetry for a center located at the center of an octagon, you do not need to show the rotational symmetry for the centers of any of the other octagons in the tessellation.

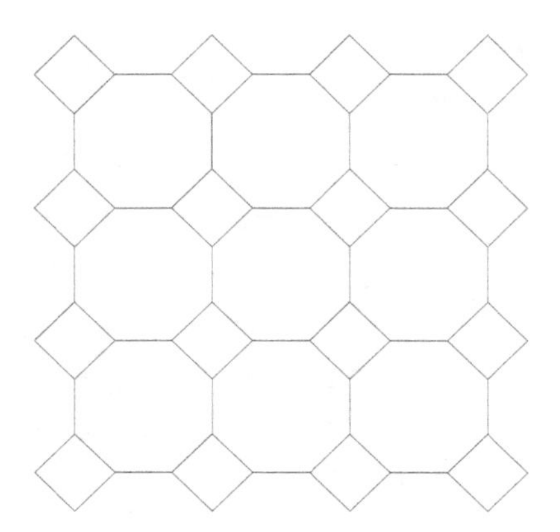

Chapter 7

Symmetry in Three Dimensions

7.1 Polyhedra

The three-dimensional version of a polygon is a polyhedron. The plural of polyhedron is polyhedra. A polyhedron is a three-dimensional figure made up of polygons joined at their edges so that together the polygons bound an inside space. Therefore, there is an "inside" and an "outside" to a polyhedron. The word derives from the Greek word polys (many) plus the hedra (seat or base) [25]. The polygons that make up the polyhedron are called the faces of the polyhedron. The edges of the polygons that make up the polyhedron are the edges of the polyhedron. A vertex of the polyhedron is a vertex of a polygon that makes up the polyhedron. An example of a polyhedron is given below:

A polyhedron is convex if for any two points in or on the polyhedron, the line segment between those two points stays entirely within the polyhedron. The above example of a polyhedron is a convex polyhedron.

If a polyhedron is not convex, it is called nonconvex. The example on the following page is nonconvex, as you can see from the picture to the right. Point A and Point B are both on the polyhedron, but the line segment $\overline{AB}$ does not stay inside the polyhedron– for instance, point C on the line segment is not inside the polyhedron.

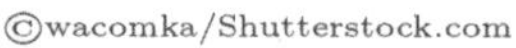
©wacomka/Shutterstock.com

©wacomka/Shutterstock.com edited by Amy Wangsness Wehe

The object below is called a tetrahedron, and is also an example of a triangular pyramid. A tetrahedron is made up of four equilateral triangles. If you were to set a tetrahedron on a table, the triangle it is sitting on would be its "base." When on its base, the rest of the edges all meet at a point at the top of the polyhedron; this point is called the apex of the pyramid. A pyramid is a three-dimensional object with a polygon as its base, one triangle face for each edge of the base, and these triangle faces meet at one vertex (called an apex) on top of the figure. In the case of a tetrahedron, each of the faces are identical, so it does not matter which one is chosen to be the base of the figure.

©Peter Hermes Furian/Shutterstock.com

Another example of a pyramid is given below. In this case, the base of the figure is a square.

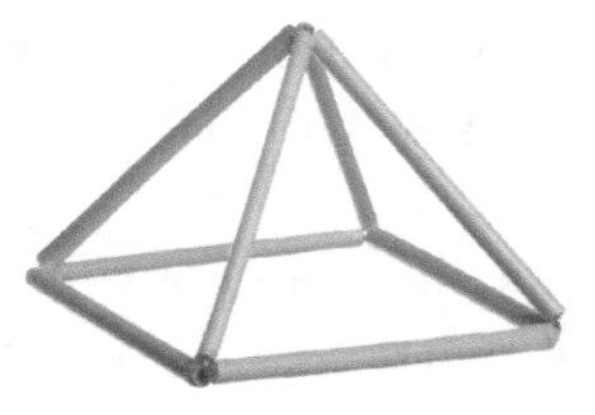

©Andresjs83/Shutterstock.com

A prism is a polyhedron with two congruent polygonal bases (sometimes referred to as faces) and a rectangular face for each side of the base polygon. The polygonal bases are parallel to each other, and the rectangles are perpendicular to the base faces. The rectangles share two edges with the base faces and the other two edges with other rectangles.

An example of a prism is given below. This example is specifically called a "triangular prism" because the base polygon of the prism is a triangle [23].

Let F be the number of faces, E be the number of edges, and V be the number of vertices of the polyhedron. Count the number of edges, the number of faces, and the number of vertices in the two pyramids on the previous page. Now compute the following:

$$F + V$$

and

$$E + 2$$

That is, add the number of faces to the number of vertices. Then take the number of edges and add 2. What do you notice about these two numbers? Now do the same thing for the cube below.

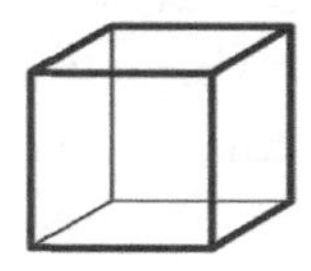

Did the same thing happen? The two numbers you found should be equal to each other. This relationship is called Euler's formula, and is usually written in the following way: $F + V = E + 2$. It is a good way to check whether or not you have correctly counted the number of faces, edges, and vertices of an object. If Euler's formula does not hold with the numbers you found, then you counted incorrectly.

You find a vertex figure of a polyhedron the same way you find the vertex figure of a tessellation. First, find a vertex of the polyhedron. Then, find all of the edges coming out of that vertex. Next, find the midpoint of each of those edges. Finally, connect the midpoints of the edges. That is the vertex figure of the polyhedron. Officially, the definition is the following: the vertex figure of a polyhedron is a polygon whose vertices are the midpoints of the polygonal edges emanating from (coming out of) a vertex of the polyhedron.

The degree of a vertex is the number of edges coming out of a vertex. Every vertex of a tetrahedron, shown below, is a vertex of degree three because each vertex has three edges emanating from (coming out of) it.

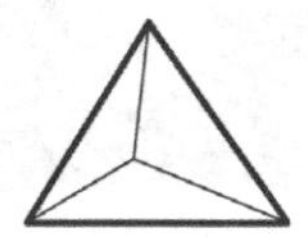

There is a very special set of polyhedra called the Platonic Solids. The platonic solids are regular polyhedra in the same way that regular tessellations were regular. Platonic Solids are polyhedra that are made up of copies of one regular polygon and have vertex figures of one size and shape. The Platonic Solids are pictured below. The names of the clear polyhedra below from left to right are tetrahedron, cube, octahedron, dodecahedron, and icosahedron..

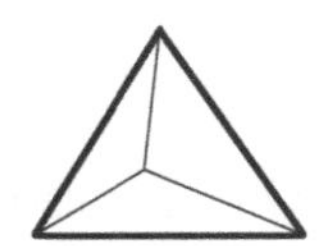 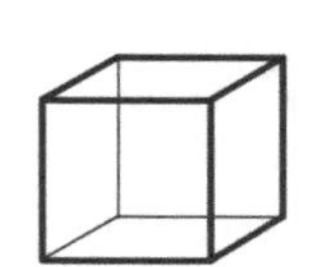 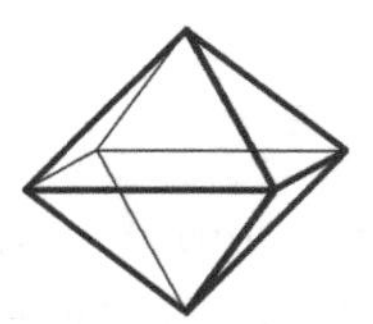 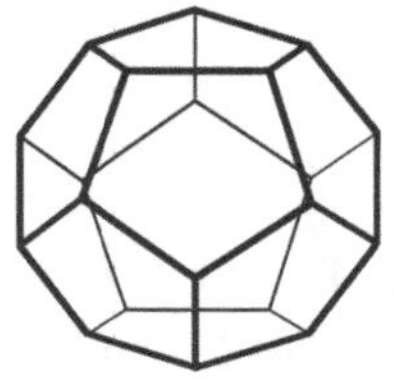 

In Section 6.3, we talked about dual tessellations. We will now talk about dual polyhedra. You find the dual of a polyhedron using the same method you use to find the dual of a tessellation. First, draw a point in the center of each face of the polyhedron (just like you would do with the prototiles of a tessellation). Then, connect two centers with a line segment if their faces share an edge; these line segments become the edges of the dual polyhedron.

For example, consider the cube. The first picture below is the cube itself. The second picture shows the vertices of the cube in black and the centers of the faces of the cube in blue. The line segments connecting the blue vertices are the edges of the dual polyhedron. What is the name of that polyhedron? This polyhedron is the dual of a cube.

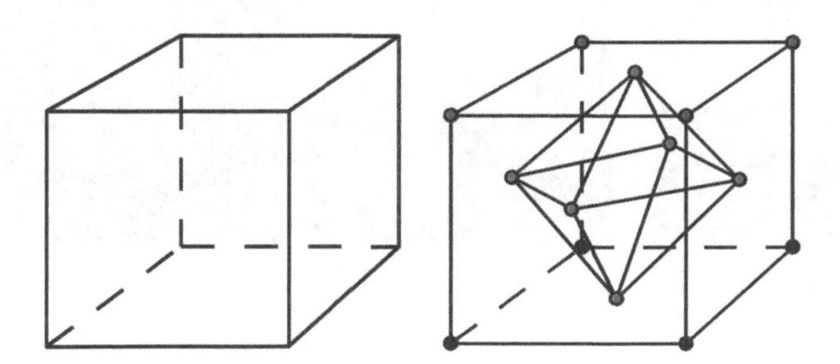

We have now talked about regular polyhedra (Platonic Solids). Just as there are semi-regular tessellations (Section 6.3), there are also semi-regular polyhedra. A polyhedron is a semi-regular polyhedron if its faces consist of more than one type of regular polygon and the vertex figures of the polyhedron are regular polygons of the same size and shape.

In fact, there are 13 semi-regular polyhedra, which are also known as Archemedean solids. These polyhedra can be found at `http://numb3rs.wolfram.com/406/`. You can also do an internet search for "semiregular polyhedra"

There are many more polyhedra. Just as there are tessellations that are neither regular nor semi-regular (see Section 6.3), there are polyhedra that are neither regular nor semi-regular polyhedra. Some of these are convex polyhedra, such as the following [6] [23]:

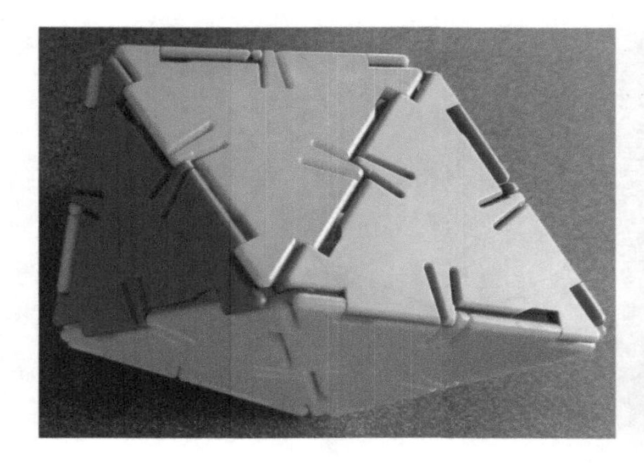

There are also many nonconvex polyhedra, such as those shown below. Nonconvex polyhedra are by definition neither regular nor semi-regular: both of these definitions require the polyhedron to be convex [22] [18].

Nonconvex "Star"

Nonconvex "Bridge"

Many more images of polyhedra can be found with an internet image search of "polyhedra." If you search for "rotating polydra," you can find some images of polyhedra that turn, so you can see the entire figure.

In the next section, we will discuss the symmetries of polyhedra.

Understanding the Text:

1. State the number of faces, vertices, and edges of the octahedron pictured on the following page. The figure has four triangles meeting at a point at the top and four triangles meeting at a point at the bottom; the cross-section of the middle of the figure is a square. Does Euler's formula hold for the octahedron?

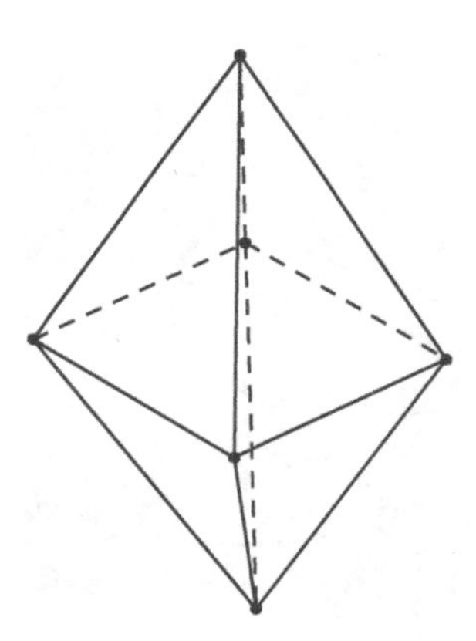

2. The dual of a cube is an octahedron, found earlier in this section. Now find the dual of the octahedron, pictured on the following page.

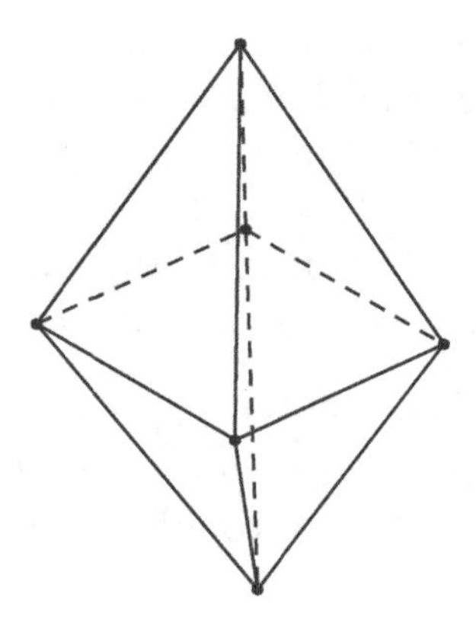

What do you notice about these two dual pairs?

3. Find and state the duals of the rest of the Platonic Solids. Polyhedron can be duals of themselves [11] [14].

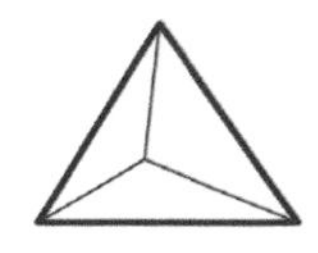

Tetrahedron

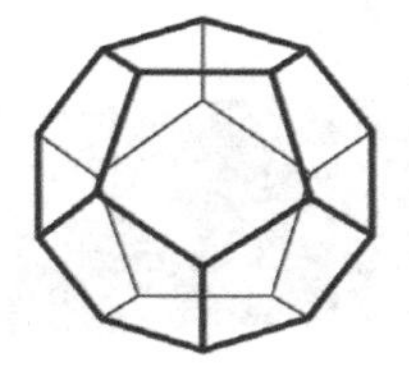

Dodecahedron

Icosahedron

4. State the degrees of the vertices of a cube, shown below.

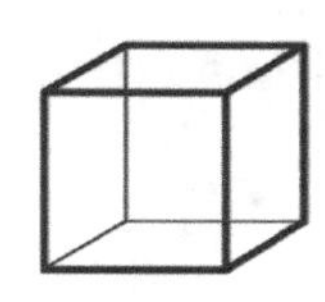

7.2 Symmetry of Polyhedra

We discussed two-dimensional symmetry in Chapter 7. In particular, we discussed the symmetry of polygons and other figures in Section 6.1. In a two-dimensional figure, rotational symmetry is determined by a point and reflectional symmetry is determined by a line. In three dimensions, rotational symmetry is determined by a line and reflectional symmetry is determined by a plane. Just like the dimension of the figure increases by one, the objects specifying symmetry also increase by one dimension.

Consider the planet earth. It is often said that the planet earth rotates on its "axis." If the earth is a sphere, and we think of it excluding any of its features (like mountains and valleys), then the earth has symmetry at any degree turn around its axis. This includes not only 1° and 2°, but also 1.2345° and $\frac{3}{1111}^{\circ}$. Just like for a circle, it is impossible to name every degree at which a sphere has symmetry. There are also many lines of rotational symmetry through a sphere– any line through the center of a sphere is an axis (line) of rotational symmetry.

Similarly, to divide a sphere into two equal parts, you would have to cut a plane through it. Picture putting a sheet of steel through the center of a sphere; as long as the sheet goes through the center of the sphere, you will divide the sphere into two equal halves, as illustrated in the picture below. Imagine that the plane (in red-orange) goes through the center of the sphere below. The sphere would be divided into two identical hemispheres.

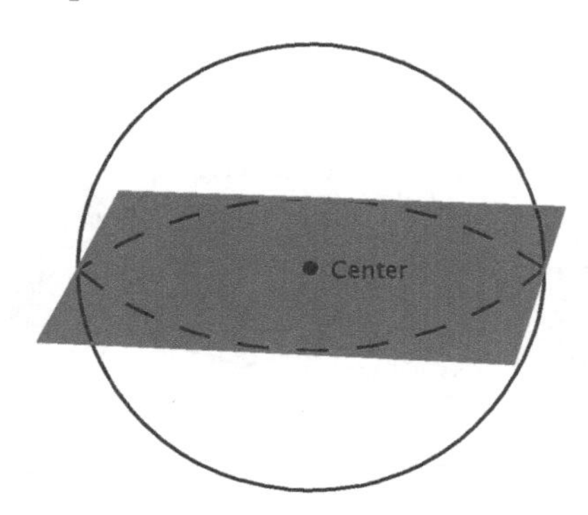

This idea works similarly for polyhedra. An axis of rotational symmetry is a line through the object so that, if the object is rotated around that line, the object looks the same as the original object at two or more different positions (counting the original position). For instance, consider the tetrahedron, discussed in the last section and pictured below. The plural of axis is "axes."

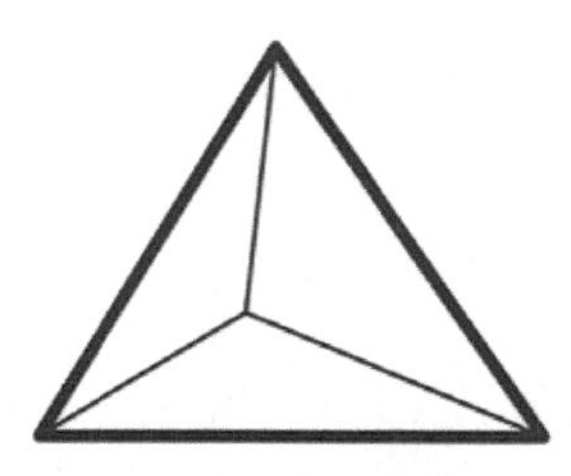

One axis of rotational symmetry goes through one of the vertices and through the center of the face opposite that vertex, as shown in the picture below.

©Peter Hermes Furian/Shutterstock.com edited by Amy Wangsness Wehe

Now, how many places does the tetrahedron look the same when you rotate it around the axis shown above? Imagine you begin with one of the vertices of the base face (the face the tetrahedron is sitting on) toward you. That is the 1st position where the tetrahedron looks the same. Now imagine you spin the tetrahedron until another vertex of the base face is toward you. That is the 2nd position. Finally, rotate it until the 3rd vertex is facing you. That is the 3rd position where it looks the same. If you rotate it one more time, to the next vertex, you will end up where you started, and you already counted that one. Therefore, there are three places where the tetrahedron looks the same when rotated around the axis pictured above. Because there are three places where the tetrahedron looks the same when rotated around this axis, we say this axis is an "order 3 axis," or an "axis of order 3."

With polyhedra, we tend to talk about the order of an axis instead of using the degree measure where the object looks the same. The order of an axis is the number of places an object looks the same when rotated around the axis. An axis of order 3 has symmetries at 0°, 120°, and 240°, but we would just write "order 3 axis," instead.

How many order 3 axes does the tetrahedron have? Look for other axes of symmetry that are analogous to the axis we already found. Our axis goes through a vertex and out the opposite face. Try another vertex of the tetrahedron. If you imagine a line drawn through that vertex and out the opposite face, is that line also an axis of rotational symmetry? Are there three places where the tetrahedron looks the same when rotated around that axis? Because the tetrahedron is a highly regular object, there is an axis of order 3 through each vertex of the tetrahedron. Since there are four vertices of the tetrahedron, the tetrahedron has four vertices of order 3.

When searching for the axes of rotational symmetry in an object, look at centers of faces, vertices, and midpoints of edges. If there is an axis of rotational symmetry through an edge, it is a rotational axis of order 2. That is because the edge has to land on itself, which will only happen if it is rotated 180°.

Can you find an axis of rotational symmetry of the tetrahedron other than the one shown above? What is the order of its axis? Since we have already looked at vertices and faces, we should look at the edges of the tetrahedron. The tetrahedron is pictured again below.

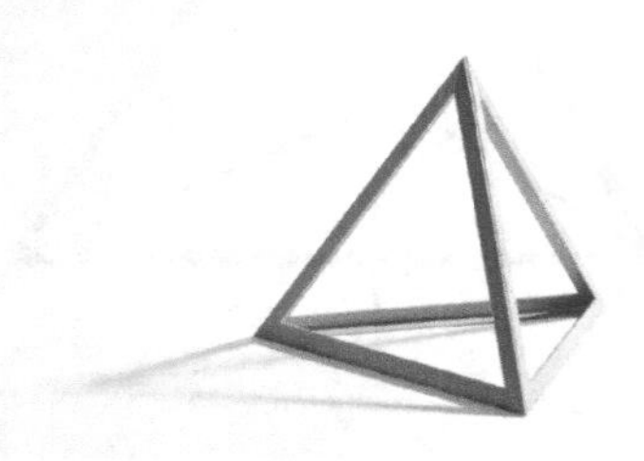

A line through the midpoint of an edge of the tetrahedron and the center of the tetrahedron will pass through the midpoint of the opposite edge. As argued above, an axis of symmetry going through the midpoint of an edge would only have rotational symmetry if it has 180° symmetry. If you rotate the tetrahedron around the axis going through the midpoint of an edge and the midpoint of the edge opposite that edge 180°, does the tetrahedron look the same? Yes, it does. Therefore, an axis going through the midpoint of an edge and the midpoint of the edge opposite that edge is an axis of order 2. How many order 2 axes does a tetrahedron have? If you simply count the edges of the tetrahedron, you will over count the number of axes because each axis goes through the midpoints of two edges. Since there are 6 edges in a tetrahedron, there must be half as many axes of rotational symmetry of order 2, because you counted each axis twice. Therefore, there are $\frac{6}{2} = 3$ axis of a tetrahedron of order 2.

Polyhedra that are not regular polyhedra will not be quite as symmetric. There may be axes through some vertices that have order 5 axes going through them and some vertices with axes going through them that are order 3 axes in the same polyhedron. Be careful when you count axes of rotational symmetry.

One example of a plane of reflection (or a plane of symmetry) in a tetrahedron is a plane that goes through one edge and splits two faces in half. In the picture on the following page, a plane of symmetry of the tetrahedron would be the plane that goes through the red line segments that have been drawn into the picture.

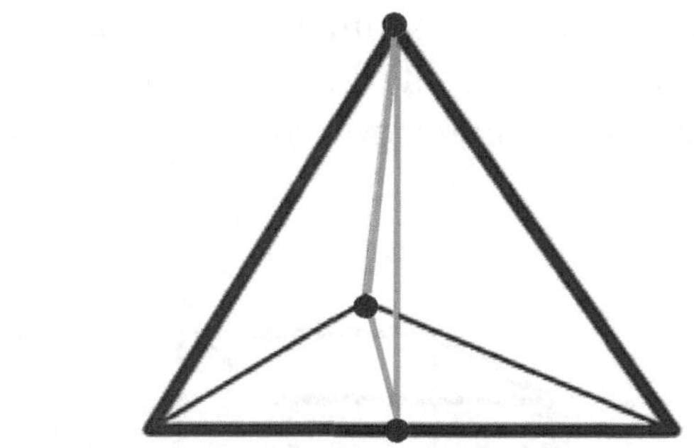

Understanding the Text:

1. Describe the different types of axes of rotational symmetry (one axis of each order) of the cube pictured below and state the number of axes of each type [8].

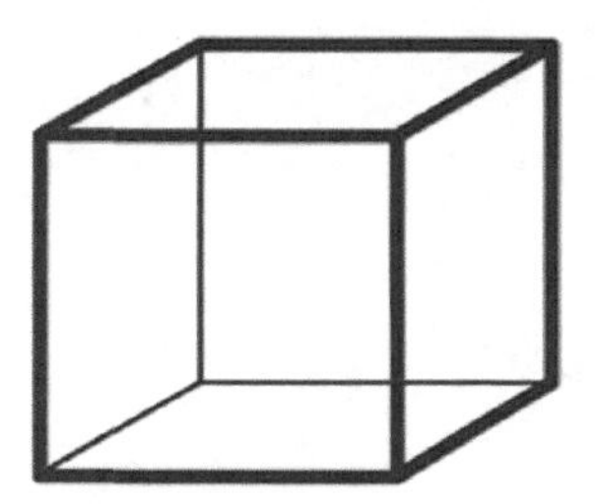

2. Describe the different types of planes of reflectional symmetry of the cube pictured below. "Different types" means planes that are not analogous to each other. That is, if one plane goes through only faces and another plane goes through both edges and faces, those planes are different [8].

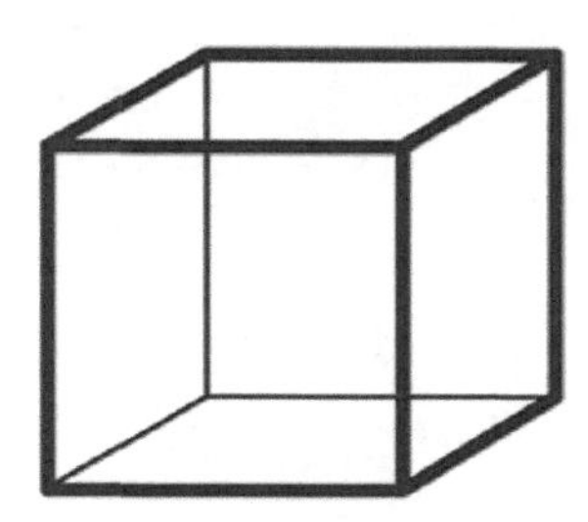

3. How many total planes of reflectional symmetry of a cube are there? Here we count analogous planes separately. That is, if there are two planes that go only through faces and three planes that go through both edges and faces, there are a total of five planes [8].

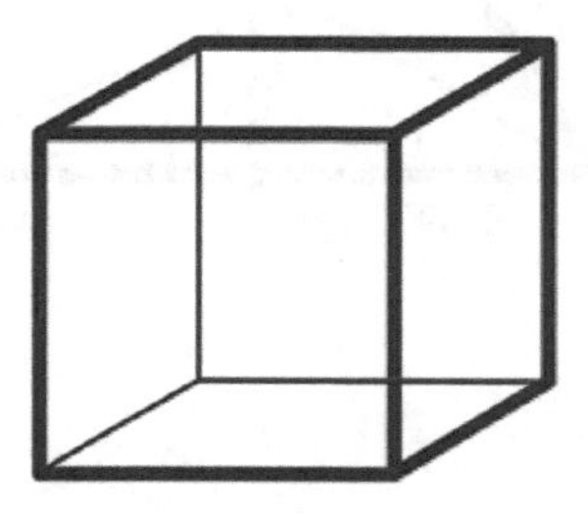

Chapter 8

Measurement

8.1 Introduction to Measurement

Measurements are created when they are needed, using what is available in that place and time. Early units were based on hands, feet, sticks, rocks, and any other number of things. Ideally, these measurements could be easily recreated and/or standardized, so that the measurement would never change. Many times, however, the measurement varies slightly from person to person, or is circular in its design.

According to legend, in a seaside village on a distant foreign shore, the village kept time by the cannon that a local military base fired at exactly noon the top of a nearby hill. The cannon fire divided morning from afternoon and was used to set up meetings. According to legend, a teenage boy wanted to know how the cannon knew to go off at exactly noon. He asked the artilleryman how he knew when to fire the cannon. He said it was his commanding officer's job to keep the most accurate watch obtainable and keep it carefully synchronized. The boy then asked the commanding officer, who showed him his finely made watch. The boy asked him how it was set. On his weekly walk into town, the commander said, he always took the same route, which carried him past the shop of the town watchmaker. He would stop and synchronize his watch with the large and venerable clock in the watchmaker's window. Then the youth asked the watchmaker how he set the large clock in the window. "I set it by the noonday cannon!" he said [28].

Clearly, this is not a reliable way to keep a measurement. Each measurement must have a standard, by which all copies of that measurement can be compared. The official measurement for a meter, for instance, has been located in Paris, France. It has been carefully kept under particular conditions of humidity and temperature, so that it does not vary. In recent times, however, scientists are working on defining each of the seven International System of Measurements (SI) units in terms of fundamental physical constants or atomic properties. In 1983, the meter was redefined in terms of the speed of light. In this way, any country can realize the meter, provided it has the technology. The International Prototype of the Meter is now a historical curiosity. It remains in the vault in France today [28].

The United States uses a combination of U.S. customary units, like inches, feet, pounds, square feet, etc, and the International System of Measurements (SI), like centimeters, kilometers, kilograms, liters, etc. There have been several movements in the United States toward "metrication," which would be toward using only SI Units for everything, but it has not taken hold. This is mainly because people in the United States do not have an intuitive sense of how long something is in meters, how warm (or cold) something is in Celsius, or how much something is in liters.

This is partially because schools have focused on conversions between units, and not on understanding the actual measurements. It is also because students do not see these units at home, because their parents do not have a feel for the units and because most things you buy in the store are not in metric units. More items are being sold in stores, however, with both U.S. customary and metric measurements. We also have 2-Liter soda bottles. Most things packaged in liters or grams, however, have not caught on in the United States.

The fact that we do not always use the same set of measurements has caused problems. For instance, in 1999, NASA lost a Mars orbiter because two groups working on the Mars orbiter team used two different sets of measurements [17]. The United States is one of the only three countries in the world that has not adopted the SI as their official system of measurements. The other two countries are Myanmar and Liberia [5].

Fortunately, for those teachers who want good real-life examples of fractions, we are still using the U.S. customary units to a great extent. Unfortunately, to those students who are not a fan of fractions, the U.S. customary units are full of fractions. We use cups for cooking ($\frac{1}{2}$ cup, $\frac{1}{3}$ cup, etc) and inches for measurement ($3\frac{5}{8}$ in, $1\frac{7}{16}$ in) etc. We also have to remember strange numbers like the fact that there are 12 in in a foot and 16 ounces in a pound. The metric system tends to use fractions in the form of decimals, since the conversion between units of measurement of the same type are factors of 10 (such as the fact that there are 100 cm in a meter or 100 g in a kilogram).

In this chapter, we will start with length and then move to volume and then weight. We will discuss a few other measurements at the end. By the way, if you are interested in getting a feel for the metric system, since the United States is still moving slowly toward it, read the metric measurement (grams and liters) on everything you buy at the supermarket– it is required to be stated on the package– and set the temperature in your car to Celsius. You know how cold it is outside– you were just out in the weather getting to your car. Seeing the temperature in Celsius associated with the weather you were just in will give you a feel for what the temperature means.

8.2 Length

8.2.1 Measurement and Conversions

In U.S. customary units, length is measured in inches, feet, yards, and miles. There are 12 in in a foot, 3 ft in a yard, and 5280 ft in a mile. Inches tend to be measured with a ruler or a tape measure. A ruler (not to scale) for inches is shown below:

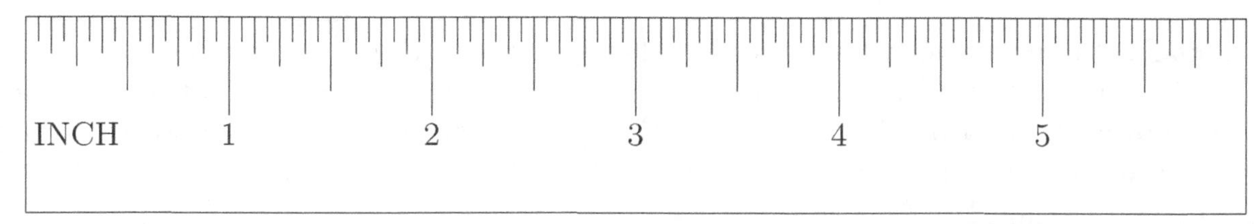

The tick marks in the ruler above are NOT tenths of an inch. They are measured using sixteenths. First of all, there are the whole inches, shown in the ruler below.

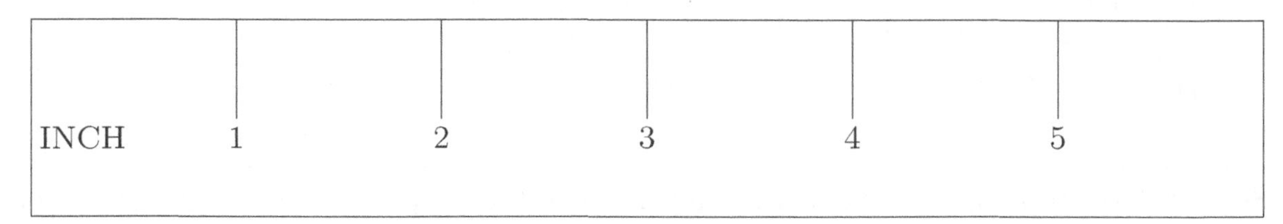

The next-size smaller tick-marks occur exactly in the middle of the inch marks. Counting spaces, there are two spaces between any two whole inch marks (one between the first inch and the next-size smaller tick-mark, and one between that tick mark and the next inch mark). Since there are two of these per inch, each of these spaces is one of two equally spaced marks per inch, and therefore measure a half-inch. Thus, each space in the ruler below is $\frac{1}{2}$ in.

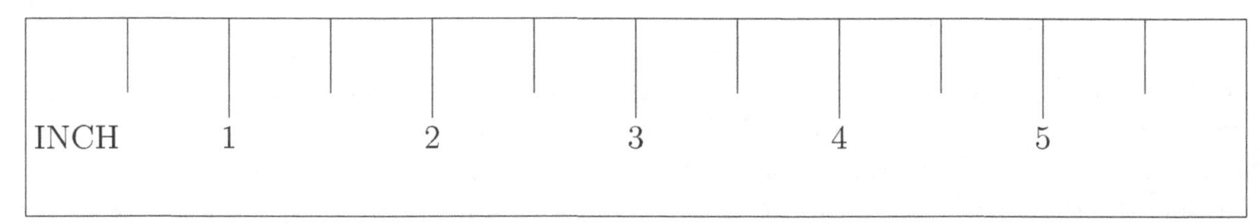

The next smaller tick marks are exactly between a half-inch mark and an inch mark. Therefore, they help divide the inch into spaces that are half the size of a half-inch; they are quarter-inch marks. These are shown in the ruler below. Notice that there are now 4 spaces between each whole-inch mark, so each space is $\frac{1}{4}$ of an inch. Note also that the half-inch marks are also on this ruler. That is because $\frac{2}{4}$ in is equal to $\frac{1}{2}$ in, so when we are counting by quarter-inches, we need to count the spaces made with the half-inch marks, also. Likewise, $\frac{4}{4}$ in = 1 in, so the spaces made with the whole-inch tick marks also need to be counted.

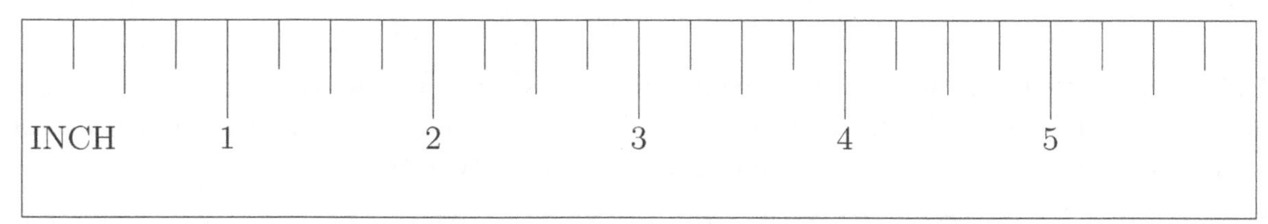

Similarly, the next-size tick marks are half-way between the tick marks in the previous ruler. They then measure half of a quarter of an inch, so they measure eighths of an inch. This is shown on the ruler below. Notice that there are now 8 spaces between any two whole-inch marks. Again, we also have to count the quarter-inch marks and the eighth-inch marks, as $\frac{2}{8} = \frac{1}{4}$ and $\frac{4}{8} = \frac{1}{2}$.

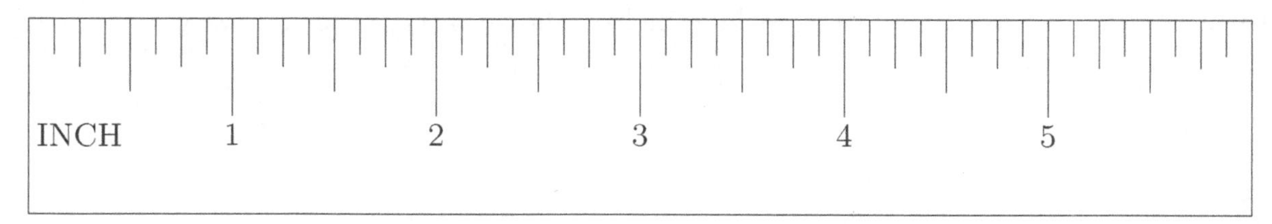

Some rulers stop there; those rulers only measure accurately to the eighth of an inch. However, some rulers continue on and measure inches to the 16th of an inch. That ruler is shown again below (it is the same ruler we started with at the beginning of this section).

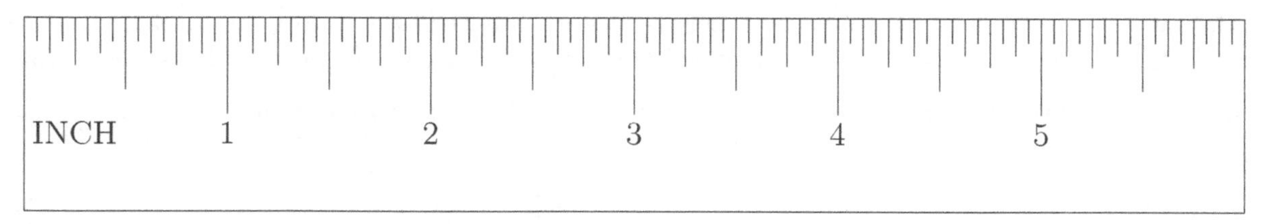

To find out how accurately your ruler measures inches, count the number of spaces between the 1-in and the 2-in tick marks on your ruler. If there are 8 spaces, the ruler measures accurately only to an eighth of an inch. If there are 16 spaces between 0 and 1 on your ruler, then the ruler measures accurately to 1/16 of an inch. I say to measure between 1 in and 2 in because some rulers have the tick-marks for 1/16 of an inch between 0 and 1, but only for 1/8 of an inch between 1 and 2. That allows you to measure things that are smaller than 1 inch more accurately, but things that are larger only accurately to the nearest eighth of an inch.

Measure the length of the line segment below using the ruler given below it. Notice by the units given on the left-hand-side of the ruler that you should be measuring in inches. Also, count the number of spaces between two inch-marks to see whether each space on the ruler below represents $\frac{1}{8}$ in or $\frac{1}{16}$ in.

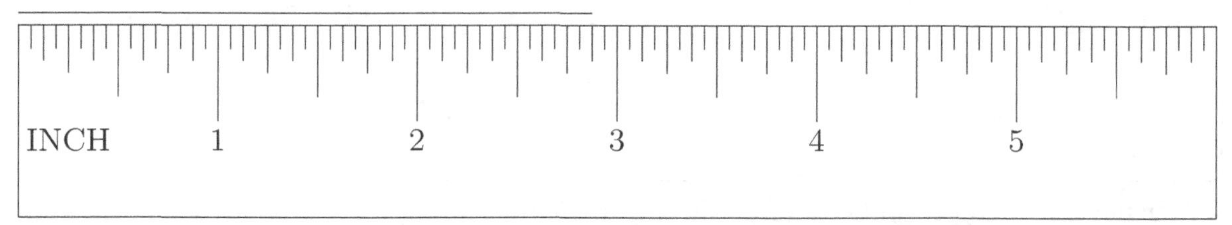

In the exercise above, you should have counted 16 spaces between each whole-inch mark, so each tick-mark on this ruler represents $\frac{1}{16}$ of an inch. The segment is at least 2 in long, because it goes past the 2 in mark. Now count the number of spaces from the 2-in mark to the end of the line segment; you should have counted 14 spaces, which means that the segment is $2\frac{14}{16}$ inches long. We do not leave our answer that way, because both the numerator and the denominator of the fraction are even: $\frac{14}{16} = \frac{7}{8}$.

To measure this without reducing your fraction at the end, you can also measure in 8ths instead of 16th. When we do this, we do not count the 16ths tick-marks when counting our spaces. Instead, count only those spaces between tick-marks longer than the shortest tick-marks. If we do this, we see there are 7 spaces between the 2-inch mark and the end of the line segment, and there are 8 spaces between the 2-inch mark and the 3-inch mark. Therefore, the end of the line segment is $\frac{7}{8}$ of the distance between the 2-inch mark and the 3-inch mark. You can see this on the ruler below, which is the same as the ruler above, excluding the 16th-inch marks.

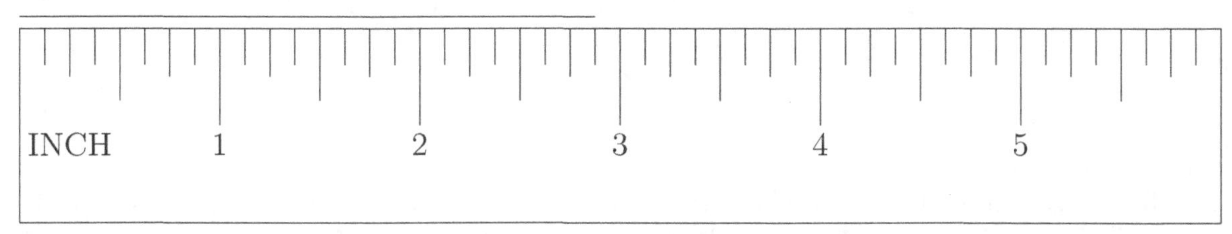

Measuring in centimeters is similar, except that each centimeter is broken down into 10 pieces. This is usually easier for us to deal with, since that matches our number system, the decimal system.

A ruler in centimeters is shown below:

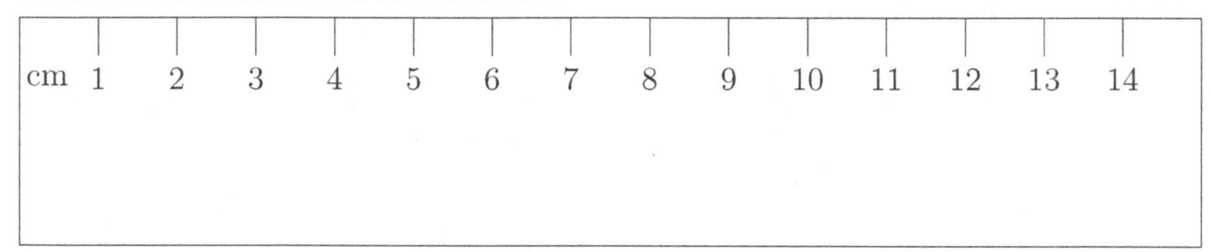

The ruler above represents 15 cm. Most centimeter rulers, however, show the tick-marks for the millimeter values. There are 10 mm in each centimeter, so we need to divide each centimeter into 10 equal spaces, as in the ruler below.

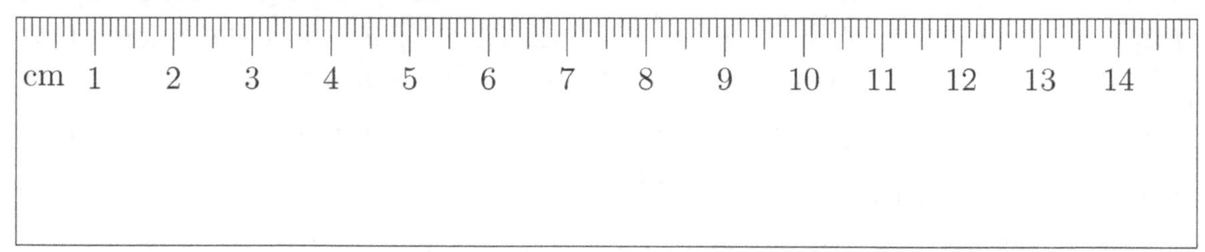

Notice that the fifth tick-mark in the ruler above is longer than the other tick-marks in between the whole-centimeter marks. This is to make it easier to count the marks; it also represents half of a centimeter.

Measure the line segment below with the given ruler, in centimeters. We know this ruler is measuring in centimeters because there is a "cm" on the left-hand-side of the ruler next to the tick-marks for this ruler.

cm 1 2 3 4 5 6 7 8 9 10 11 12 13 14

The line segment you were to measure is at least 7 cm in length, since it goes past the "7" on the ruler, which is the 7-cm mark. Counting the number of millimeters past the 7cm mark, we see that there are 5 spaces. Therefore, the line segment uses 5 of the 10 spaces before the 8-cm mark. So the line segment is $7\frac{5}{10}$ cm long. Centimeters are usually measured using decimals, but notice that if you read this out loud, you get 7 and 5 tenths centimeters; that number, written as a decimal, is 7.5 cm. Notice that this is also 7 and $\frac{1}{2}$ cm, which is consistent with the fact that the line segment ends on the fifth tick-mark, which is the half-way point between 7 cm and 8 cm.

It may also be advantageous at times to write this number in millimeters. Since there are

10 mm in each centimeter, 7 cm is the same as 70 mm. If you are not sure about this, count 10 millimeters for each of the 1cm spaces on the ruler: 10mm, 20mm, 30mm, 40mm, 50mm, 60mm, 70mm; now you are at the 7-inch mark and have counted 70mm. Next, we need to add on the additional 5 millimeters needed to get to the end of the given line segment above. Therefore, the line segment is 75mm long.

You may have noticed that in the previous paragraph we have 75mm, and in the paragraph before that one, we had 7.5cm, and both of these measurements are measuring the length of the same line segment. Yes, in fact, 7.5cm = 75mm, and yes, since there are 10 mm in each centimeter and we use the decimal system (so there are 10 numbers in each place-value before moving to the next place value), converting from centimeters to millimeters is equivalent to moving the decimal point one place. Which direction do you move the decimal place? Your answer should have more millimeters than centimeters, because it takes more millimeters to measure the same length (because millimeters are smaller).

Now, let us look at some other measurements of length:

In U.S. customary units, feet are used to measure length when inches are too unwieldy, but when you are not talking about a very large distance. One foot is equivalent to 12 inches. It is meant to be approximately the length of an actual foot, but it is quite a bit bigger than most people's feet.

In U.S. customary units, yards are used to measure the distance when "stepping off" distance and for lengths on a football field. Yards are estimated by steps; one large step by the average person is about 1 yd. Of course, this varies quite a bit from person to person. It is also about the length from the tip of your finger to the shoulder of your other arm; this is how we used to measure the lengths of embroidery floss required for one friendship bracelet when I was a kid. A yard is also equal to 3 ft or 36 in.

The mile is a U.S. customary unit used in the United States for large distances, such as distance between towns and the altitude of planes in the air. A mile is 5280 ft.

In SI units, the type of objects one would use inches to measure in U.S. Customary Units would be measured in centimeters or millimeters. The type of objects one would use feet or yards to measure in U.S. Customary Units would be measured in meters in SI units. Objects one measures in miles in U.S. Customary Units are measured in kilometers in SI units. In fact, although the people of the United States tend to use U.S. Customary Measurements, the standard for length in the United States is based on the yard, and the yard is defined based on the standard for the meter (1 yd = 0.9144 m) as of 1959 [26].

All SI units use the same prefixes for the different measurements. The base unit does not have a prefix. In the case of length, a "meter" is the base unit. There is an SI unit of measurement for each factor of 10 of that base measurement. Each of these units are a prefix of the base unit. The following table gives some of these prefixes.

Prefix	Symbol	Equivalency	Number of Base Units in Each
tera-	T	0.000,000,000,001	1,000,000,000,000
giga-	G	0.000,000,001	1,000,000,000
mega-	M	0.000,001	1,000,000
kilo-	k	0.001	1000
hecto-	h	0.01	100
deka-	da	.1	10
base unit		1 base unit	1 base unit
deci-	d	10	0.1
centi-	c	100	0.01
milli-	m	1000	0.001
micro-	μ	1,000,000	0.000,001
nano-	n	1,000,000,000	0.000,000,001
pico-	p	1,000,000,000,000	0.000,000,000,001

Use the column "Equivalency" to make conversions between the units. The units in this column are given numbers representing equal quantity. For instance, 0.001 km is equal to 100 cm, which are both equal to 1 m. The numbers in the "Number of Base Units in Each" column gives the number of meters that equal one such unit. For instance, there are 100 m in a hectometer, and there are 0.001 m in a millimeter. The word "kilo" means 1000, and so "kilometer" is 1000 m. The prefix "centi" is 100, or in this case 100th. There is 100th of a meter in a centimeter; or you can think of it as 100 cm in a meter.

All you need to do a unit conversion are two equivalent quantities. That is, you need to know a number of one unit that is equivalent to a number of the other unit. For instance:

How many kilometers is equivalent to 500 m?

Here we have kilometers and meters. We want to know a number of kilometers that are equal to a certain number of meters. From the "Equivalency" columns of the table above, we see that 1 m = 0.001 km.

If 1 m is equal to 0.001 km, then 500 m must be 500 times as many kilometers as 0.001 km. Therefore, 500 m = $0.001 \times 500 = 0.5$ km.

In talking ourselves through the problem, we used proportions to figure out the answer: 1 m is to 0.001 km as 500 m is to how many kilometers? Or, written out as a proportion, that statement is:

$$\frac{1 \text{ m}}{0.001 \text{ km}} = \frac{500 \text{ m}}{x \text{ km}}$$

You can solve this by by solving the equation for x. One way to do this is to work as if you are finding a common denominator. That is, the common denominator is $0.001x$. Then the first fraction becomes $\frac{x}{0.001x}$ and the second becomes $\frac{500 \times 0.001}{0.001x}$. Now the fractions have a common denominator, and the numerators can be compared directly.

$$\frac{x}{0.001x} = \frac{500 \times 0.001}{0.001x} \text{ so } x = 500 \times 0.001 = 0.5 \text{ km}$$

There is yet another way to think about this problem. That is using Unit Cancellation. Begin with what you know, which is 500 m. Then you want to convert the units into kilometers.

$$500 \text{ m} = \frac{500 \cancel{\text{m}}}{1}\left(\frac{0.001 \text{ km}}{1 \cancel{\text{m}}}\right) = \frac{0.5}{1} \text{ km} = 0.5 \text{ km}$$

Notice here that the factor you multiplied by initially (in parentheses) is equal to 1 because the length 0.001 km is equal to the length 1 m. That is 0.001 km = 1 m, so the numerator and the denominator of that factor are equal in amounts to each other. Multiplying by 1 does not change the original quantity. Therefore, the resulting 0.5 km is equal to the original 500 m. Also notice that we divided by meters so the meters' units would cancel, and multiplied by kilometers, because that is what we wanted to end up with.

Sometimes we do not have a direct equivalency between two units available to us, but we have an equivalency from each unit to a third unit. For instance, we know that there are 12 in in a foot, and we know that there are 5280 ft in a mile, but we do not know, off-hand, how many inches are in a mile. However, we can find out how many miles are equal to 5000 in by using the equivalency between inches and feet and between feet and miles.

$$5000 \text{ in} = \frac{5000 \cancel{\text{in}}}{1}\left(\frac{1 \cancel{\text{ft}}}{12 \cancel{\text{in}}}\right)\left(\frac{1\text{mi}}{5280 \cancel{\text{ft}}}\right) = 0.0789 \text{ mi}$$

The same method can be used if you have a chain of equivalencies between two units you would like to convert between. For instance, there are 2.54 cm in 1 in, there are 12 in in a foot, and there are 100 cm in 1 m. How many feet are equivalent to 5 m?

$$5 \text{ m} = \frac{5 \cancel{\text{m}}}{1}\left(\frac{100 \cancel{\text{cm}}}{1 \cancel{\text{m}}}\right)\left(\frac{1 \cancel{\text{in}}}{2.54 \cancel{\text{cm}}}\right)\left(\frac{1 \text{ ft}}{12 \cancel{\text{in}}}\right) = 16.4042 \text{ ft}$$

Perhaps you noticed above that you were given an equivalency between SI units and U.S. Customary measurements. Here are a couple more. Just for practice, check these. Since you know that there are 2.54 cm in 1 in, you have a way to get from any metric measurement to any U.S. Customary measurement through a chain of equivalencies. Try it and see if you get the equivalencies below:

$$\begin{aligned} 1 \text{ in} &= 0.0254 \text{ m} \\ 1 \text{ mi} &= 1.609344 \text{ km} \end{aligned}$$

All of the equivalencies above are of the form 1 "something" is equal to some quantity of "something else." This is not necessary. For instance, you can use the equivalency 5000 in = 0.0789 mi, which we found in an example above, to convert between inches and miles. For instance, how many miles is 8000 in?

$$8000 \text{ in} = \frac{8000 \cancel{\text{in}}}{1}\left(\frac{0.0789 \text{ mi}}{5000 \cancel{\text{in}}}\right) = 0.12624 \text{ mi}$$

This works because 0.0789 mi measures the same distance as 5000 in, so the quantity in parentheses equals 1. Multiplying by 1 does not change the quantity you started with.

Next, we will talk about measurement in two dimensions (area). Most of the principles discussed in this section will apply for area as well. We just have to be careful that we are always working in the correct dimension.

Understanding the Text:

1. Find the length of the line to the right in inches. ______________________________
2. If the distance between two cities is 400 km, how far apart are they in miles? (1 mi = 1.609344 km).
3. You are going to be in a wedding in Europe, and the bride tells you to buy silver shoes with 4 cm heels for the event. When you go out looking for appropriate shoes, you need to be able to tell the shoe employees how high the heels should be in inches. What should you tell them? (1 cm = 0.3937 in).

8.2.2 Perimeter and Circumference

The perimeter of a simple polygon is a measure of the outer boundary of the object. For instance, the perimeter of the pentagon below is the sum of the lengths of its sides.

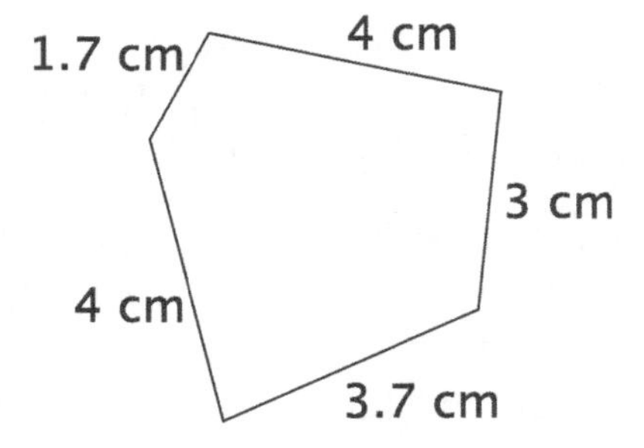

Therefore, the perimeter of the pentagon is $1.7 + 4 + 3.7 + 3 + 4 = 16.4$ cm. Note that this is also the measure of the boundary of the pentagon. That is, you would need 16.4 cm of thread to *exactly* cover the boundary of the pentagon.

You can measure the perimeter of a circle as well, but the perimeter of a circle has a special name. It is called the circumference of a circle. Before we find the circumference of a circle, we start with a couple of definitions we will need. The diameter of a circle is the length of the longest line segment in a circle with endpoints on the circle. The diameter of a circle can also be defined as the length of a line segment with endpoints on the circle, which passes through the center of the circle. The actual line segment in question is also called a diameter of a circle.

The radius of a circle is the length of a line segment with one endpoint on the center of the circle and the other endpoint on the circle. The radius of a circle is $\frac{1}{2}$ the diameter of a circle.

We find the circumference of a circle by using a relationship that was already known at the time of the ancient Egyptians and ancient Babylonians: that the circumference of a circle is a little more than 3 times the length of the diameter of the circle [29]. Throughout history, people tried to get closer and closer to the actual value of this ratio, $\frac{C}{d}$. Eventually, it was discovered that the ratio is an irrational number. This number is now known as π.

Using the ratio, we see that $\frac{C}{d} = \pi$, where C is the circumference of the circle and d is the diameter of the circle. Using a little algebra (multiplying both sides by d), the circumference of a circle is $C = \pi d$.

To illustrate, we find the circumference of the circle on the following page. In many cases, at least in textbooks, the information you are given about a circle is the radius of the circle. The formula we have for the circumference of a circle has the diameter, rather than the radius. We can change the formula to involve the radius by using the fact that $d = 2r$ where $d =$ diameter of the circle and $r =$ radius of the circle. This means that the formula for the circumference of a circle, which is $C = \pi d$, is the same as $C = \pi \times 2r$ where the d is replaced by $2r$. It is standard to put the numbers ahead of the variables in a formula, so we write the formula as $C = 2\pi r$.

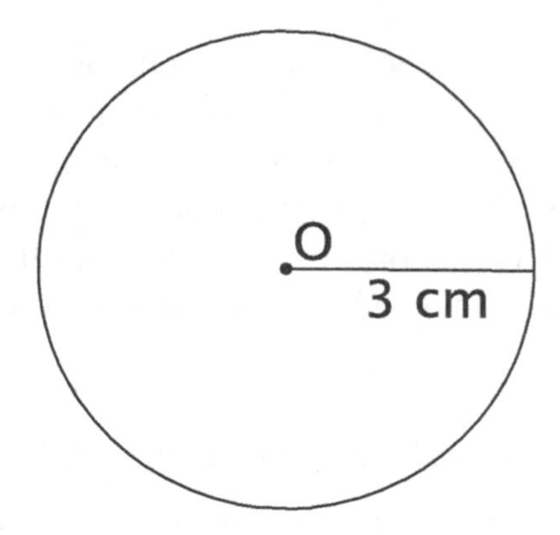

Using the formula $C = \pi d = 2\pi r$, the circumference of this circle is $2\pi \times 3 = 6\pi$ cm $\approx$ 18.85 cm.

8.3 Area

8.3.1 Area Formulae

Consider the picture on the left below of a square (not to scale) that is 5 inches by 5 inches. The area of that square is 25 square inches, which you can see by dividing the square up into 25 squares that are each 1 in by 1 in. The picture of the square divided into 25 square inches is shown to the right.

5 in
5 in | 25 sq in

1 in 1 in 1 in 1 in 1 in
1 in
1 in
1 in
1 in
1 in

Therefore, the area of a square can be found by taking the length of the square and multiplying it by the width of the square. In fact, this technique can be used for any rectangle. A square is just a special case of a rectangle.

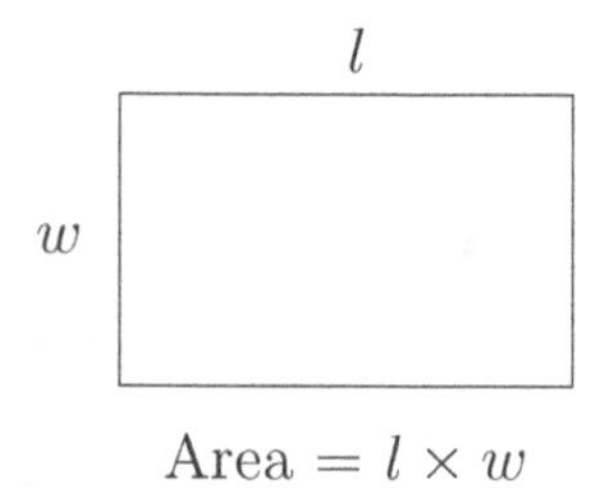

As discussed above, this works because the area is equal to the number of 1×1 unit squares that will fit inside the given rectangle. For example, to find the area of the rectangle on the left below, we draw as many 1×1 unit squares as will fit in the rectangle and add them up. Since there are three rows of eight squares in the rectangle, this is equivalent to multiplying 8 by 3.

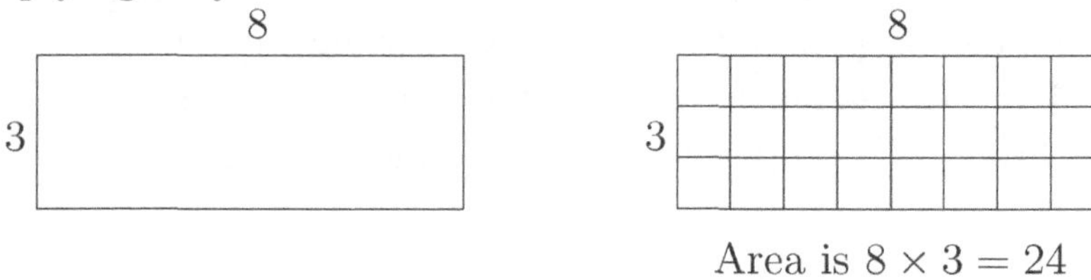

Area is $8 \times 3 = 24$

Another way to represent the area of a rectangle is "length of the base times the height.," where the base of the rectangle is the length of one of its sides and the height of the rectangle is the length of a side intersecting the base.

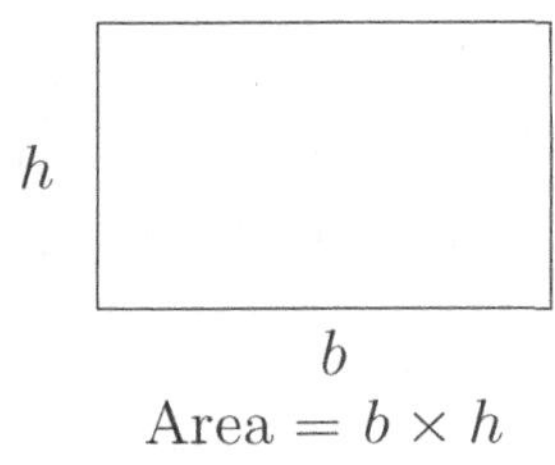

Area $= b \times h$

The area of a parallelogram is also equivalent to the length of the base times the height of the parallelogram; however, you have to be careful about the height. The height of a parallelogram is the length of a line segment from a vertex of the parallelogram to the opposite side (called the base) that is perpendicular to the base of the parallelogram.

For example, consider the parallelogram below.

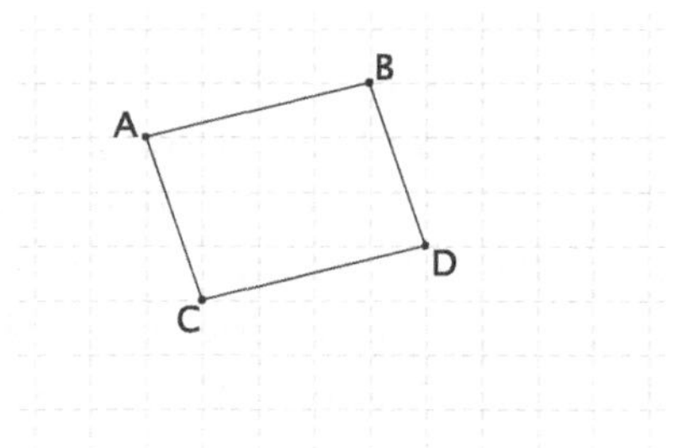

You can tell this is a parallelogram because of the following: to get from Point A to Point B, you go to the right 4 units and up 1 unit, and you do the same to get from Point C to Point D; to get from Point A to Point C, you go down 3 units and right 1 unit, and you do the same to get from Point B to Point D. This tells us the following:

- The lines containing sides that are opposite each other have the same slope.
- The sides of the parallelogram opposite each other have the same length.

To find the area of the parallelogram, we move pieces of the parallelogram around until we get a rectangle. First, create a line perpendicular to the base through a point on the opposite side. In the case of the parallelogram above, this means drawing a line perpendicular to $\overleftrightarrow{CD}$ through Point B. Then doing the same thing with the side opposite the base of the parallelogram; in this case, $\overleftrightarrow{AB}$ through C. This is shown in the picture below.

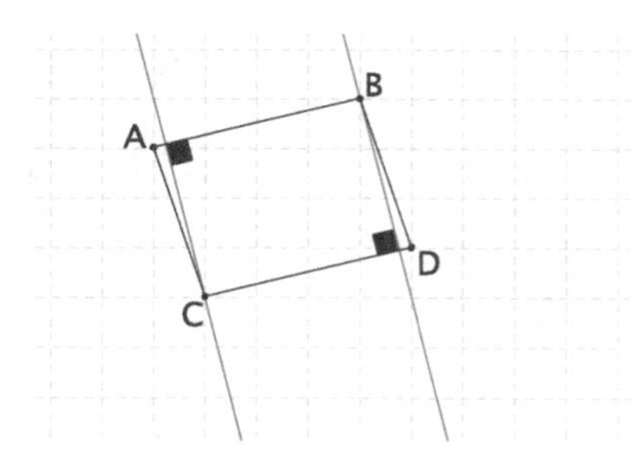

Now copy one of the resulting triangles to the other side of the parallelogram. The resulting rectangle (in red below) has the same area as the original parallelogram.

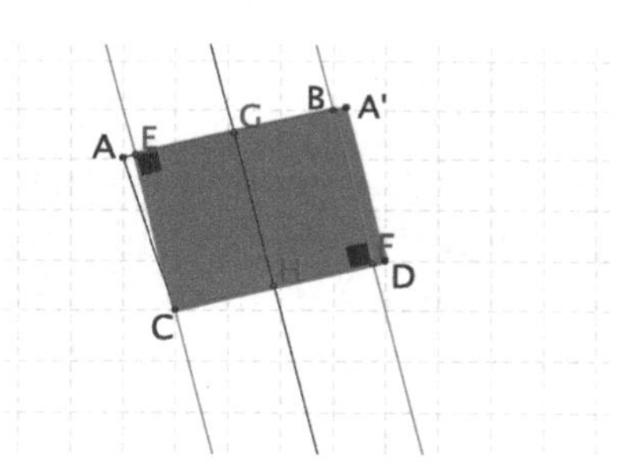

This can be done with any parallelogram. Notice that the height of the rectangle is the length of the perpendicular line between the base of the parallelogram and the side opposite the base, and that the width of the rectangle is equivalent (the same length, but in a slightly different location) as the base of the parallelogram.

In this way, we see that the area of the parallelogram is the base times the height, where the height is the perpendicular distance between the base and the side opposite the base.

For another example, the area of the parallelogram below is the length of the base times the perpendicular height.

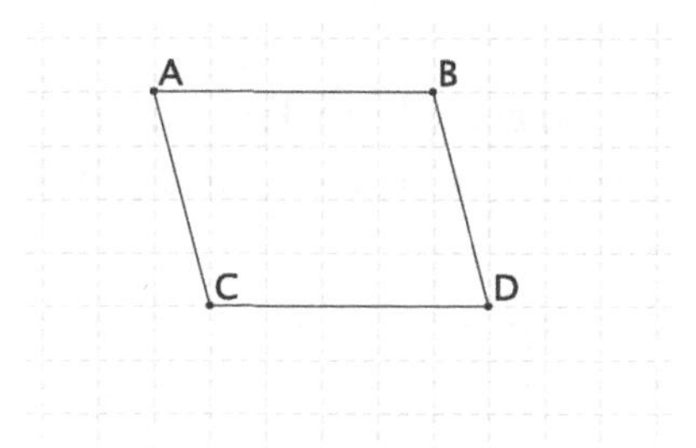

In this case, the length of the base is 5, because there are 5 unit lengths between Point C and Point D. The perpendicular height is 4, which you can see using the perpendicular line from $\overrightarrow{CD}$ to the point B. As you can see in the figure below, there are 4 spaces between Point B and Point E.

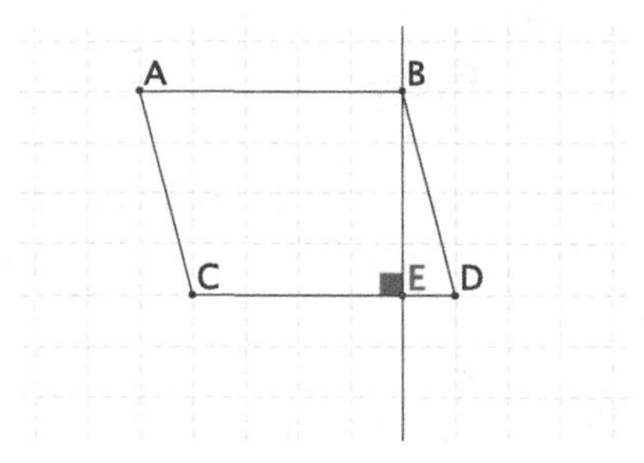

Therefore, the area of this parallelogram is $5 \times 4 = 20$ units. If you move ΔBED over to the other side of the parallelogram, you can see the rectangle with the equivalent area.

The area of a triangle is half of the area of a parallelogram with the same base and height as the triangle's base and height. You can see examples of this in the illustrations below. Note that a rectangle is a special case of a parallelogram.

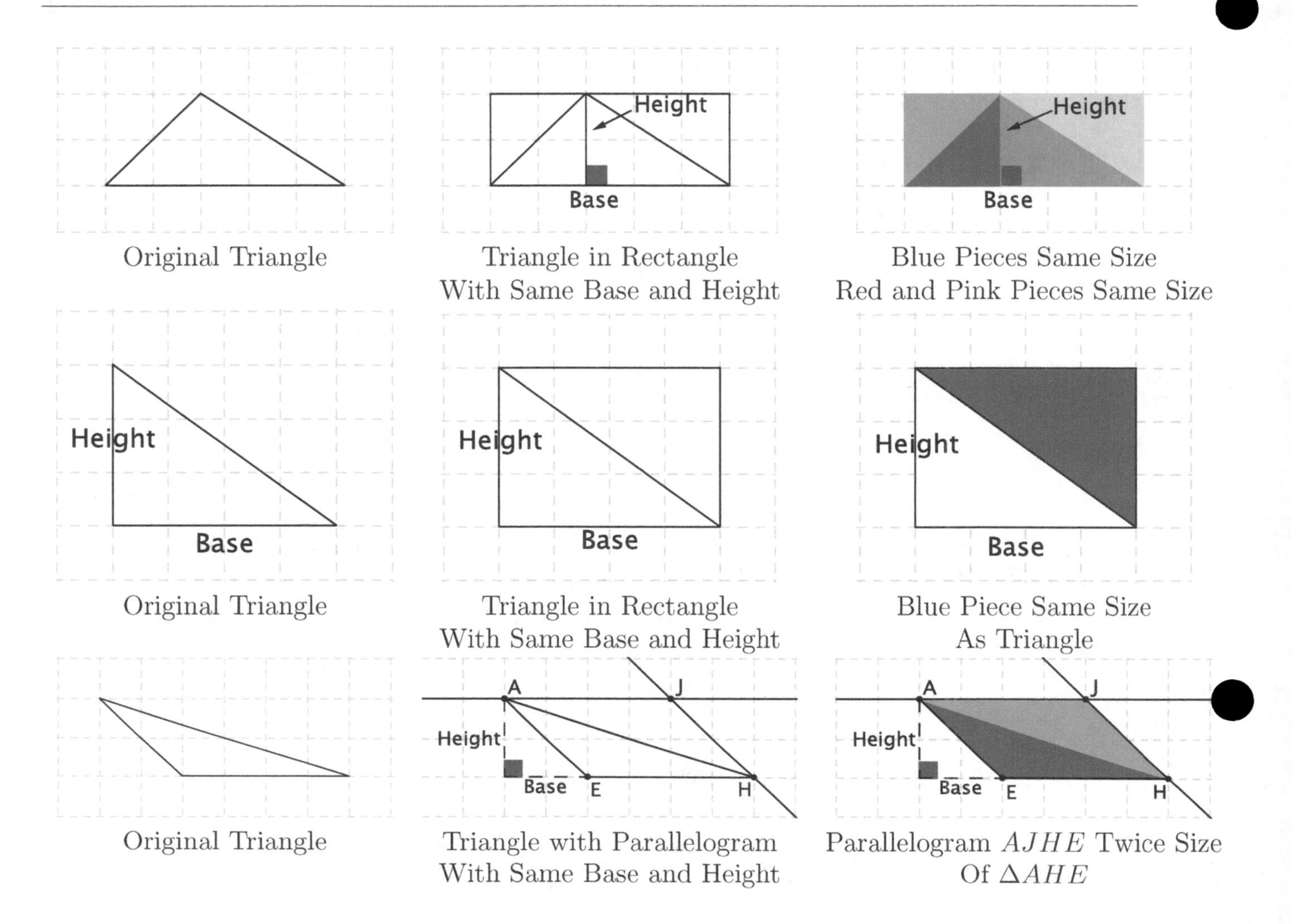

Original Triangle — Triangle in Rectangle With Same Base and Height — Blue Pieces Same Size Red and Pink Pieces Same Size

Original Triangle — Triangle in Rectangle With Same Base and Height — Blue Piece Same Size As Triangle

Original Triangle — Triangle with Parallelogram With Same Base and Height — Parallelogram $AJHE$ Twice Size Of ΔAHE

The original triangle is on the left. The center picture in each row shows the parallelogram (or rectangle) with length and height the same as the triangle's base and height. In the third picture, the portions of the parallelogram that are equal in size are highlighted. In this way, you can see that the area inside the triangle is the same as the rest of the area in the parallelogram. Therefore, the parallelogram is twice as big as the triangle. This can be done with any triangle; if you are not yet convinced, try some more triangles on your own.

We next find the area of a trapezoid. Recall that a trapezoid is a figure with at least one pair of parallel sides. Consider the examples of trapezoids below.

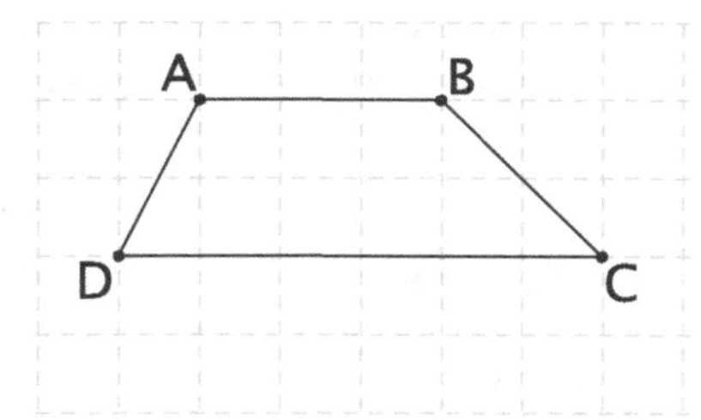

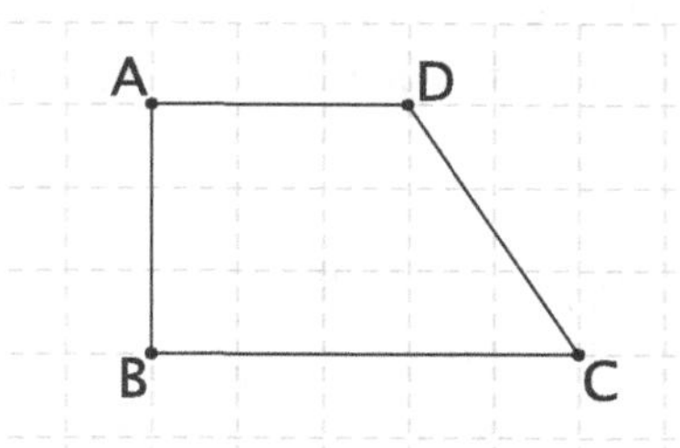

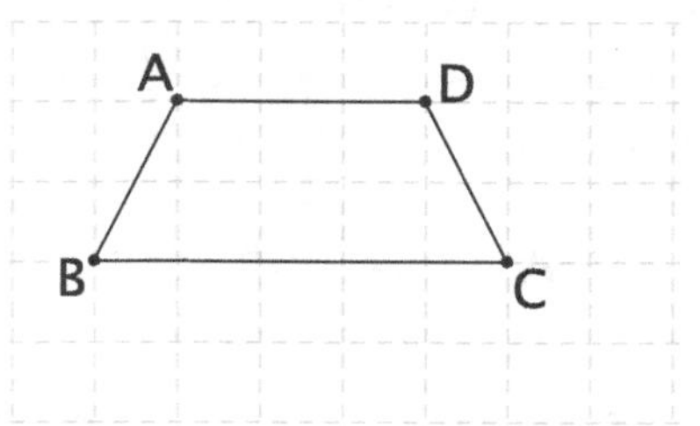

To find the area of these trapezoids, break them down into shapes such as rectangles and triangles. Then you can find the area of the trapezoid by finding the areas of the component pieces. An efficient way to do this for each of these trapezoids is shown below:

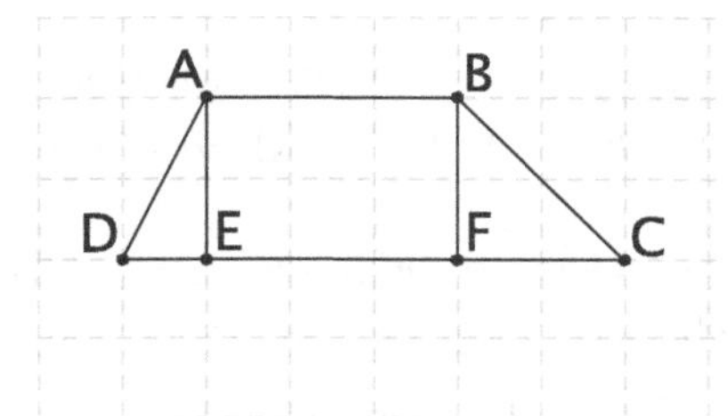

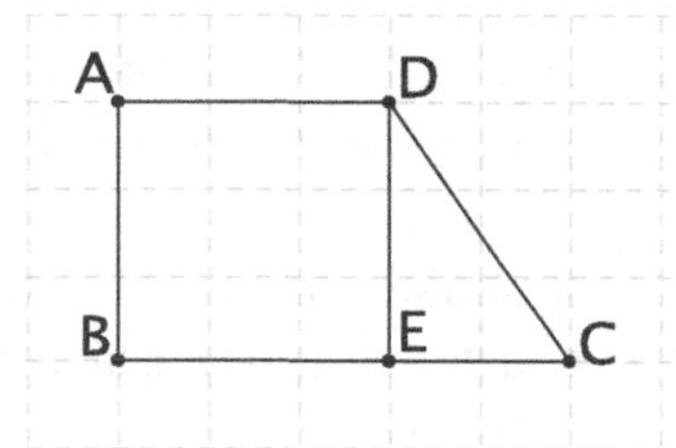

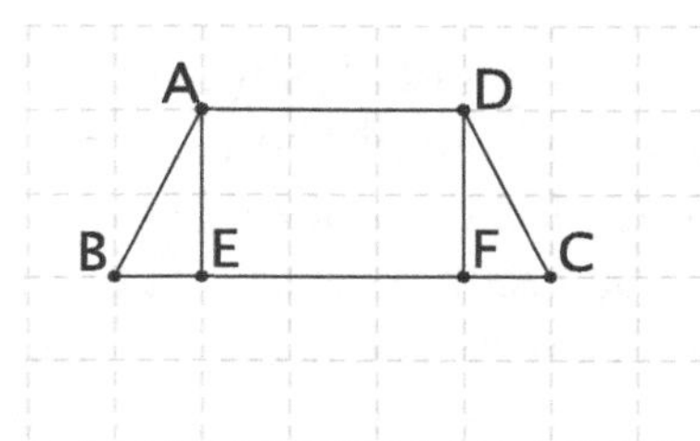

Now each of the trapezoids is a conglomeration of triangles and rectangles, which we have already discussed above. Find the area of each of the pieces and then add them together to find the area of the trapezoids.

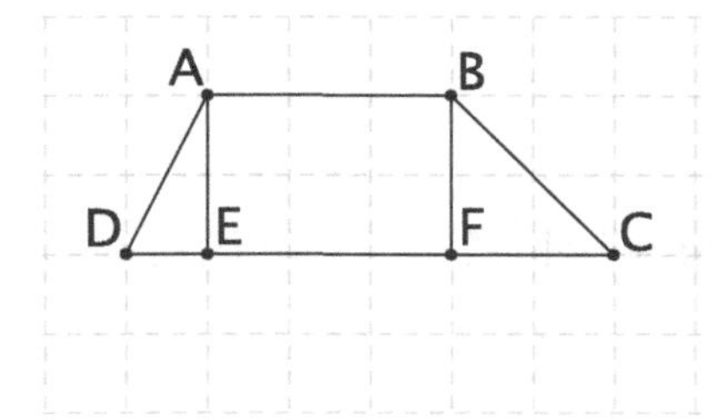

$\frac{1}{2}(1)(2) + (3)(2) + \frac{1}{2}(2)(2) =$
$1 + 6 + 2 =$ 9 units

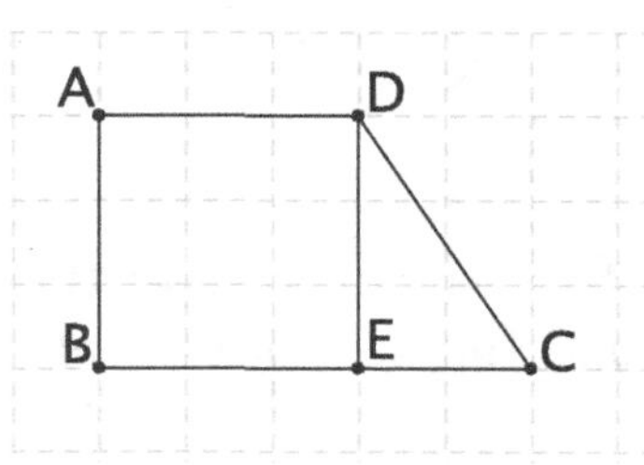

$(3)(3) + \frac{1}{2}(2)(3) =$
$9 + 3 = 12$ units

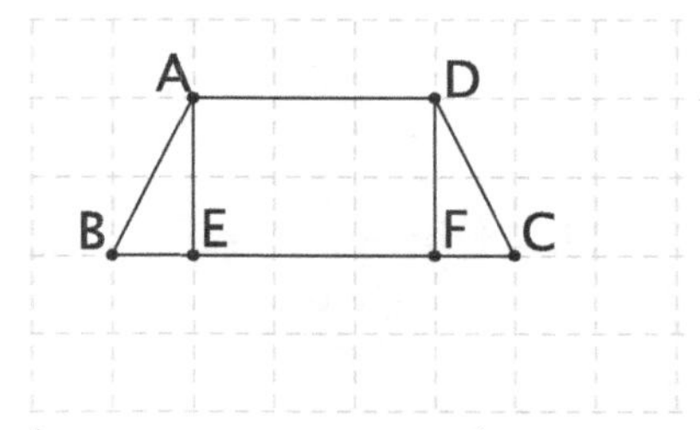

$\frac{1}{2}(1)(2) + (2)(3) + \frac{1}{2}(1)(2) =$
$1 + 6 + 1 = 8$ units

If you do a web search for the formula of the area of a trapezoid, you will likely find something like $A_{\text{Trapezoid}} = \frac{a+b}{2}h$. What does that mean? The a and the b stand for the lengths of two parallel sides of the trapezoid, and h represents the length of a perpendicular line segment between those two sides. In fact, $\frac{a+b}{2}$ represents the average of the lengths of the two parallel sides. In order to find the average of a group of things, you add them all up and divide by the number of things; in this case, the two "things" are the lengths of the parallel sides. Use this formula on the trapezoids above and see if you get the same answer as you did before.

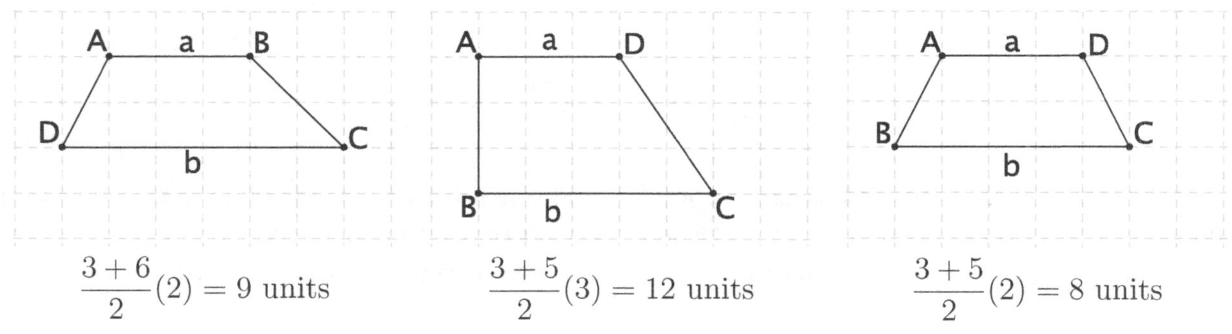

$\frac{3+6}{2}(2) = 9$ units $\quad \frac{3+5}{2}(3) = 12$ units $\quad \frac{3+5}{2}(2) = 8$ units

The idea of this formula is that we can think of the trapezoid as a rectangle whose base is the average of the length of the bases and whose height is the same as the perpendicular height of the trapezoid.

You can find the area of many two-dimensional objects by breaking the object down into the pieces whose area you know. This is true, in particular, if the object is drawn on a grid.

For instance, we do not have a formula for finding the area of the polygon below.

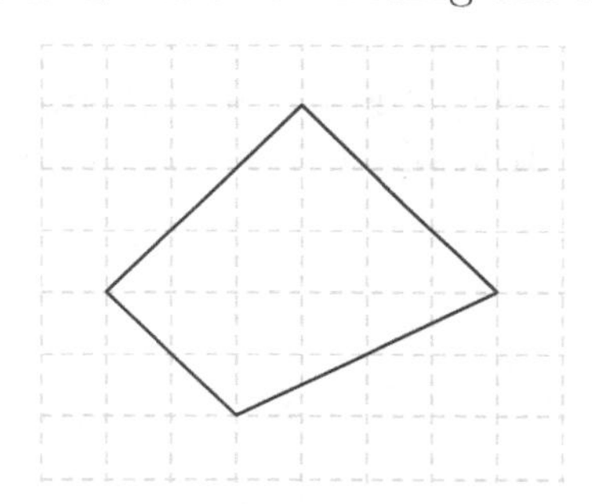

This figure can be divided into two triangles, as in the figure below. We then compute the areas of these two triangles to find the area of the entire figure.

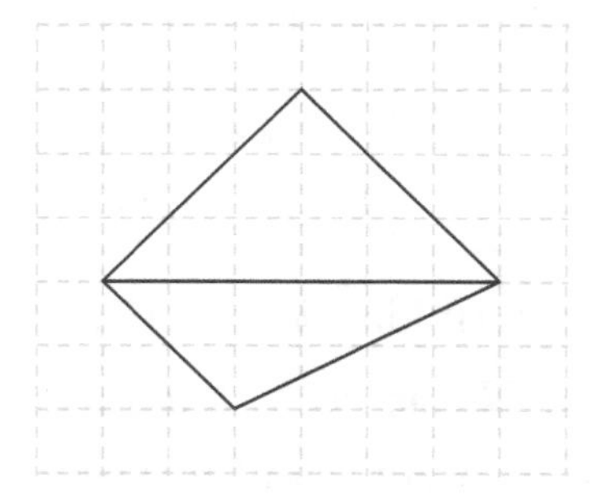

Now all we need to do is count spaces for the base and height of the triangles, and then use the formula $\frac{1}{2}($ base $\times$ height$)$ to compute the area of each triangle. Finally, the area of the whole figure is the sum of the areas of the two triangles.

The measurements of the base and heights of the two triangles are shown below. Both triangles have a base of 6 units. The top triangle has a height of 3 units and the bottom triangle has a height of 2 units. Notice that the height of each triangle is the length of the line segment through a vertex of the triangle and perpendicular to the opposite side. This is called the "perpendicular height" of the triangle. It is also called an altitude of the triangle.

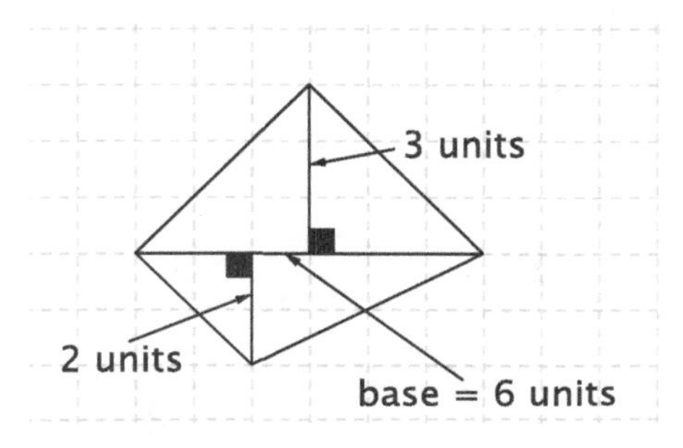

Therefore, the area of the figure is

$$\frac{1}{2}(6)(3) + \frac{1}{2}(6)(2) = 9 + 6 = 15 \text{ square units}$$

Sometimes it is easier to find the area of a larger figure and then subtract some of the area to find the area of a given figure. For instance, consider the figure below.

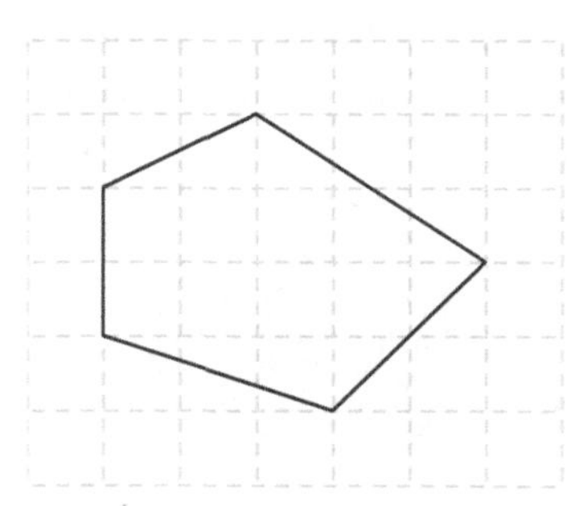

There is no obvious way to divide this figure into triangles and squares with easy to compute areas. However, we can consider this figure as part of a larger rectangle, as shown below:

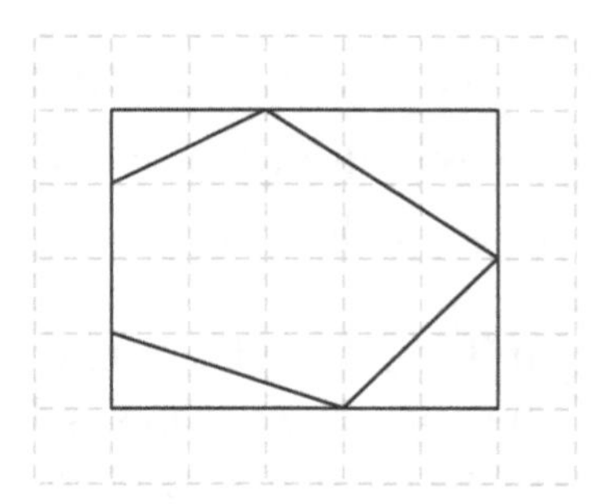

We can then find the area of the larger rectangle and subtract the areas of the triangles in its corners; what is left will be the area of the original figure. The area of the rectangle is length × width, which in this case is $(5)(4) = 20$. Next, we need to find the base and height of each of the four corner triangles:

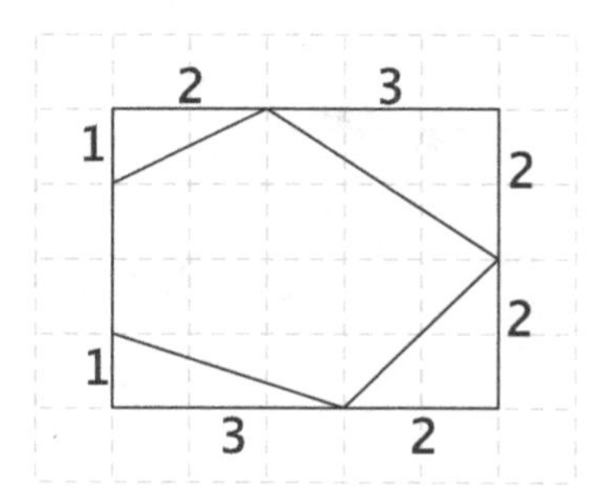

The area of the four triangles (counter-clockwise from upper left) are $\frac{1}{2}(1)(2) = 1$, $\frac{1}{2}(1)(3) = \frac{3}{2}$, $\frac{1}{2}(2)(2) = 1$, and $\frac{1}{2}(2)(3) = 3$. Therefore, the area of the original figure is $20 - (1 + \frac{3}{2} + 2 + 3) = 20 - 7\frac{1}{2} = 12\frac{1}{2}$ or 12.5 square units.

The area of a circle is $A_{circle} = \pi r^2$. One way to see this is to think of a circle created like an onion, in small layers, similar to the picture below.[29]

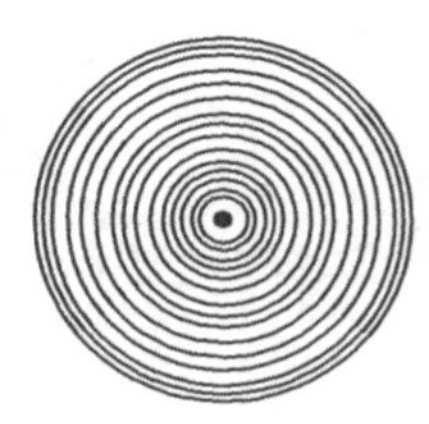

Now, in your mind, take a pair of scissors, starting from the outside of the circle and cut toward the center of the circle. Then all of the onion layers of the circle will be cut in the same place of the circle. The resulting figure would look similar to the figure below.

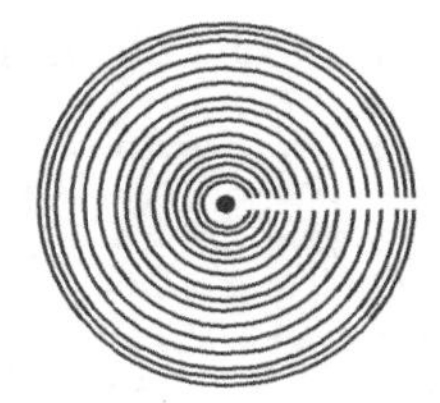

Next, stretch out each of the onion layers like string and stack them up on top of each other. The resulting picture should look something like the figure below, which is a triangle.

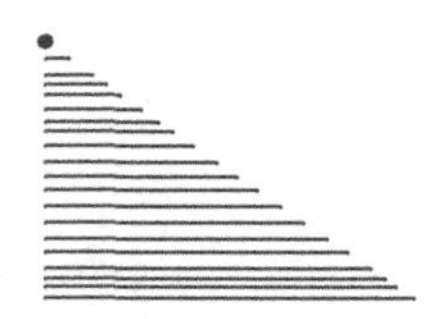

Now find the area of the triangle. The length of the base of the triangle is the length of the longest piece of string. The longest piece of string came from the outside of the circle, so the length of the base is the same as the circumference of the circle. The height of the triangle is created by stacking the pieces of string. Therefore, the height of the triangle is equivalent to the radius of the circle. Let r represent the radius of the circle. We know from Section 8.2.2 that the circumference of a circle is $C = 2\pi r$, where r is the radius of the circle. Therefore, the area of the triangle is

$$\frac{1}{2}(2\pi r)(r) = \pi r^2$$

Since the area of the triangle is the same as the area of the circle, the area of the circle is also πr^2.

If you are still having trouble imagining how the circle can become a triangle, you can view a moving illustration of this in wikipedia under "area of a disk" under the section heading "Triangle Method." This is the link: `http://en.wikipedia.org/wiki/Circle_area`.[3]

Understanding the Text:

1. The 813-square foot apartment mentioned in this section cost 780 Euros per month. If 1 USD = 0.73 Euro, how many dollars does that apartment cost per month? [15]

2. Find the circumference and area of the circle below.

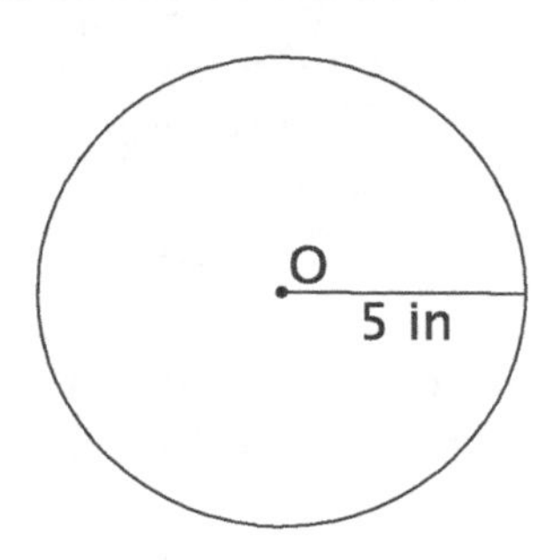

8.3.2 Area Conversions

Again, there are measurements of area in U.S. customary units and measurements in SI units (the metric system).

In U.S. customary units, there are many ways of measuring area that are used for different purposes. A few are given here.

$272\frac{1}{4}$ square feet (ft^2)	=	1 square rod (sq rd)
160 square rods (sq rd)	=	1 acre
640 acre	=	1 square mile (mi^2)
1 square mile (mi^2)	=	1 section of land
1 township	=	36 sections

These are equivalencies, and converting between them requires the same procedure used in Section 8.2.

Square feet are used to measure rooms of houses for carpet, wallpaper, and for measuring the size of a house to sell it. Square rods are used to measure land, as are acres and square miles. In the western states of the United States, counties are divided into townships. Each township is usually approximately 36 mi^2; this varies due to the local politics at the time the townships were determined, the ethics of the surveyor who determined the township lines, and the geographical restrictions of the land. [28]

Of course, area can be measured in any "square" unit associated with any length unit, such as square inches or square meters. To convert between square feet and square inches, we have to be careful. It is not the same thing as converting between feet and inches.

The equivalency between feet and inches is: 1 in = 2.54 cm. We would like to find an equivalency between square centimeters and square inches. We can do this by finding an equivalency between a 1×1 in square and the equivalent square in centimeters.

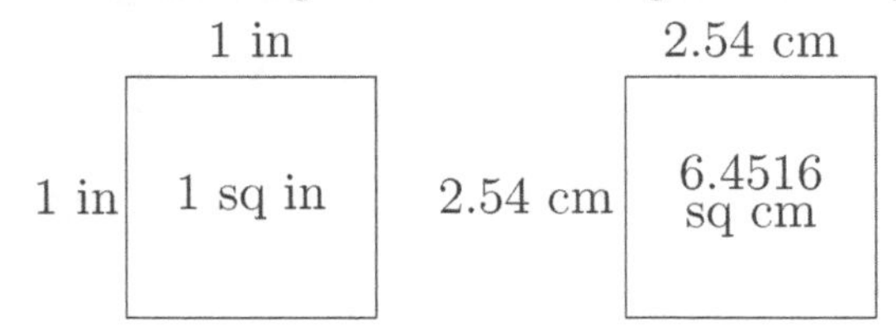

The 1-in by 1-in square on the left is equivalent to the 2.54-cm by 2.54-cm square at the right because 1 in = 2.54 cm and so the lengths of the sides of the square are the same. The area of a 1 in by 1 in square is 1 in^2. The area of a 2.54 cm by 2.54 cm square is $2.54 \times 2.54 = 6.4516\ \text{cm}^2$. We now have an equivalency between square inches and square centimeters: namely, that 1 in^2 is equal to 6.4516 cm^2.

Now, what if we want to find the equivalent of 25 in^2 in square centimeters? We use the equivalency we found above: $1\ \text{in}^2 = 6.4516\ \text{cm}^2$. Therefore, 25 in^2 is equal to

$$25\ \text{in}^2 \times \left(\frac{6.4516\ \text{cm}^2}{1\ \text{in}^2}\right) = 161.29\ \text{cm}^2.$$

Here is another example. Say you plan to study in Berlin for a semester, and you need to find an apartment there. The ads for apartments, however, show the size of the apartment in square meters. You have no feeling for how large an apartment is when it is given in square meters. However, you have some idea if it is measured in square feet. You find one apartment for 780.00 Euros per month all inclusive with 75.5 m^2 of space [15]. How big is that apartment in square feet?

First, we find an equivalency between square feet and square meters. We start with the equivalency between feet and meters, which is 1 ft = 0.3048 m. Next, we create two squares with an area of 1 ft^2 square foot to find an equivalency between square feet and square meters.

1 ft

1 ft | 1 sq ft

0.3048 m

0.3048 m | 0.09290304 sq m

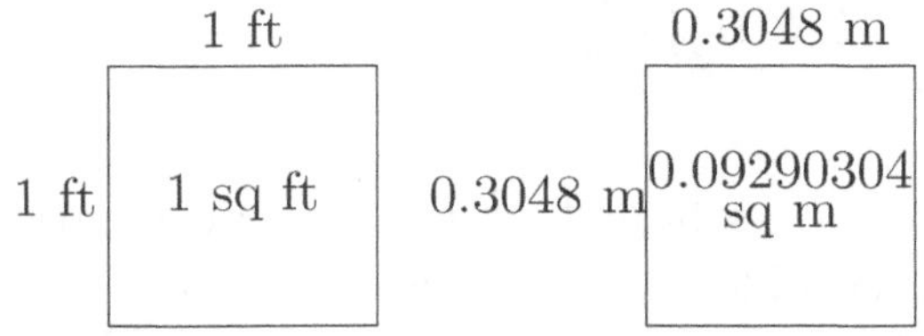

Therefore, an equivalency between square feet and square meters is $1\ \text{ft}^2 = 0.09290304\ \text{m}^2$. Finally, we use this equivalency to find the size of the apartment in square feet.

$$75.5\ \text{m}^2 \left(\frac{1\ \text{ft}^2}{0.09290304\ \text{m}^2}\right) \approx 813\ \text{ft}^2$$

Understanding the Text:

1. Another apartment in Berlin costs 670.00 Euro all inclusive per month for 99.5 m^2 of space [15].

   (a) How much does the apartment cost in dollars per month if 1 USD = 0.73 Euro?

   (b) How large is the apartment in square feet?

2. If the area of a room is 630 ft^2, how many square yards is the room? There are 3 ft in 1 yd. Notice that you will need to first find an equivalency between square feet and square yards using 1 yd squares; then you will need to use the equivalency you found to find the size of the room in square yards.

8.3.3 Surface Area

The surface area of a three-dimensional object is the total area of the outside surface of the object. You can think of the surface area as the area of the wrapping paper necessary to exactly cover the outside of a package. By exactly cover, I mean that there would be no overlapping or gaps in the wrapping paper on the surface of the package.

For instance, the surface area of a rectangular prism is the sum of the areas of all six of its faces. To illustrate, we find the surface area of the rectangular prism below.

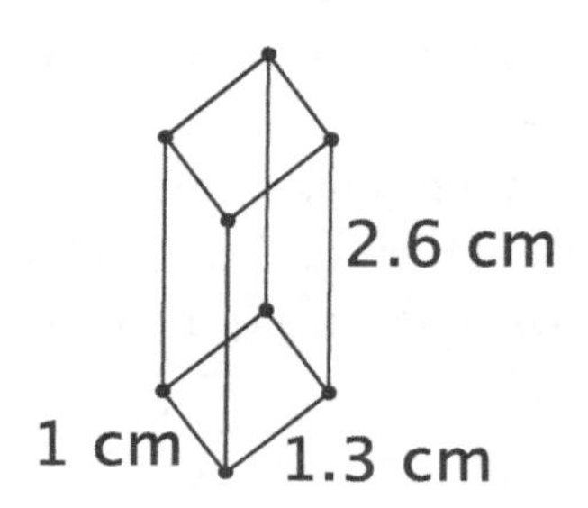

The area of a rectangle is $A = lw$ where l represents the length of the rectangle and w represents the width of the rectangle. In this case, the rectangle at the base of this rectangular prism has a length of 1 cm and a width of 1.3 cm. Therefore, the area of the base of the prism is $1 \times 1.3 = 1.3$ cm^2. In a rectangular prism, the top of the prism is the same size and shape as the base, so that part of the prism also has an area of 1.3 cm^2. One of the sides of the prism has a width of 1.3 cm and a length of 2.6 cm. Therefore, this side has an area of $1.3 \times 2.6 = 3.38$ cm^2. The side directly across from this one has the same size and shape, and so it also has an area of 3.8 cm^2. There are two more sides to this rectangular prism. They are identical in size and shape; each has a width of 1 cm and a length of 2.6 cm. Therefore, they each have an area of 2.6 cm^2.

Putting all of the information in the previous paragraph together, we see that this rectangular prism has a surface area of $1.3 + 1.3 + 3.38 + 3.38 + 2.6 + 2.6 = 14.56$ cm^2.

Next, let us look at how to write the surface area of a particular type of prism in general. For instance, consider an equilateral triangular prism; an example of such a prism is given below. Call the height of a triangle h, an edge of a triangle s (since the triangle is equilateral, all of the edges of the triangle are the same length), and the height of the prism H. [24]

The surface area of the prism is the total area of the outside of the prism. To find this, we find the area of each of the faces and add them all up. The area of an equilateral triangular base in this case is $\frac{1}{2}sh$. The bottom and top of the prism are both the same sized triangles, so we have two triangles with this area. The area of one of the rectangular faces is, in this case, sH, because the width of a rectangle is the same length as the side of a triangle, and the height of the rectangle is the same as the height of the prism. There are three rectangles of equal size in this prism. Therefore, the surface area of this prism is $\frac{1}{2}sh + \frac{1}{2}sh + sH + sH + sH$. This can also be written as $2 \times \frac{1}{2}sh + 3sH = sh + 3sH$.

To find the surface area of any figure, find the area of each of the surfaces of the figure and add them up. See Section 8.3 for help in finding the area of the various faces.

8.4 Volume

In U.S. customary units, volume is measured in cubic inches (cu. in. or in^3), cubic feet (cu. ft. or ft^3), and cubic yards (cu. yd. or yd^3). There is also a group of units for measuring the volume of liquids and one for measuring volumes of dry material.

Fluid volume is measured by fluid ounces (fl. oz.). One fluid ounce is $\frac{1}{16}$ of a U.S. pint (pt). There are 2 pints in a quart (qt), and 4 quarts in a gallon (gal) (hence the name "quart," as in a "quarter of a gallon"), and 1 cup (cp or c.) is 8 fl. oz. Some things are also measured in "barrels," but the amount a barrel contains is different depending on what you are measuring, such as beer, kerosene, or petroleum. There are also teaspoons (tsp or t.) and tablespoons (Tbsp or T.). There are 3 teaspoons in one tablespoon and 2 tablespoons in 1 fl. oz.

Dry volume is measured in pints (pt), quarts (qt), peck (pk), bushels (bu), or barrels. A dry pint (550.6105 mL) is different from a liquid pint (473.176473 mL). However, a dry quart is still 2 dry pints. A peck is 8 dry quarts. A bushel is 4 pecks, and a dry barrel is 3.281 bu.

We restate the definition of a prism here; it was originally stated in Section 7.1. A prism is a polyhedron with two congruent polygonal bases (sometimes referred to as faces) and a rectangular face for each side of the base polygon. The polygonal bases are parallel to each other, and the rectangles are perpendicular to the base faces. The rectangles share two edges with the base faces and the other two edges with other rectangles.

An example of a prism is given below. This example is specifically called a "triangular prism" because the base polygon of the prism is a triangle [23].

To find the volume of a prism, we find the area of the base of the prism and multiply it by the height of the prism. This is because a prism is basically the base shape stacked on top of itself over and over. For instance, think about a square prism; a square prism is a prism with a square base. The picture shown below is a square prism (with side length we are going to call s) divided into chunks that are one unit high [21].

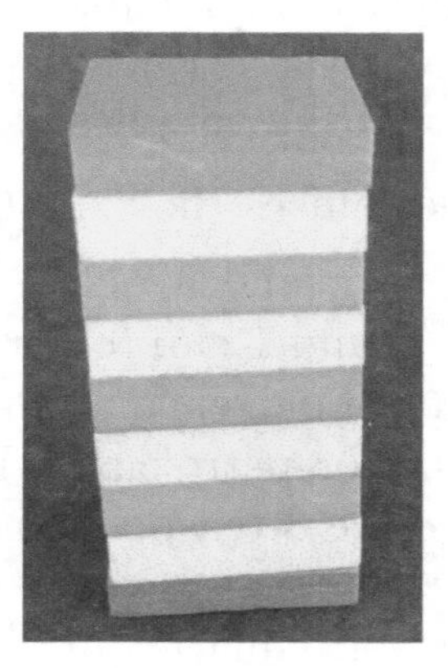

How many one-unit-high squares are required to make this particular prism? 9. Therefore, the volume of this prism is the area of the base (s^2) times the height of the prism (9

units, in this case), so the volume of this prism is $9s^2$.

In general, if you call the height of the prism h and the length of the side of a square s, the volume of a square prism is s^2h.

Similarly, we can look at a circular prism, which is also called a cylinder. The cylinder pictured on the following page is divided into 1-unit-high cylinders with a radius we will call r [9].

How many 1-unit-high cylinders are required to make the entire cylinder? In this case, it is 11. Therefore, the volume of this cylinder is $\pi r^2 \times 11$, because πr^2 is the area of a circle with radius r, and the cylinder is 11 units high.

In general, if you call the height of the prism h and the radius of a circular base r, the volume of cylinder is $\pi r^2 h$. A cylinder can be thought of as a circular prism.

A pyramid is a three-dimensional figure with a polygonal base (also sometimes referred to as a face) and a triangular face for each side of the base polygon; the triangular faces all meet at one point, called the apex, of the triangular pyramid.

Just as in Section 8.2, if you are given the equivalency between two units measuring volume, you can use that equivalency to convert between the two units of volume. However, just as in Section 8.3, if you are given an equivalency between units of length (such as centimeters or inches) and asked to convert between units that are the cubes of those lengths (such as cubic centimeters or cubic inches), you cannot use the length unit conversions directly. Instead, you need to create an equivalency between the cubic units directly.

For example, consider a cube that is 1 yd by 1 yd by 1 yd, as the cube on the left on the following page is. Such a cube would have a volume of 1 yd^3. That same cube would be how many cubic feet? Recall that there are 3 ft in 1 yd, but those are lengths; they can only be used to convert the lengths of the sides of the cubes, not to convert the volumes of the cubes themselves.

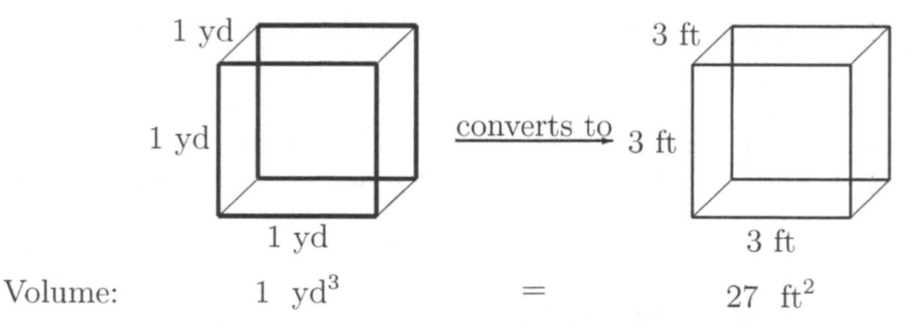

The volumes of these cubes give us an equivalency between cubic yards and cubic feet; that is, that 1 yd^3 is equivalent to $3 \times 3 \times 3 = 27$ ft^3.

For example, how many cubic yards is 54 ft^3? Use the conversion 1 yd^3 = 27 ft^3, so $54\ \text{ft}^3 \left(\frac{1\ \text{yd}^3}{27\ \text{ft}^3}\right) = \frac{54}{27}\ \text{yd}^3 = 2\ \text{yd}^3$.

8.5 Weight

The main unit used for weight in the U.S. customary units is the pound. A pound is dependent on gravity, so things weigh more on Earth than they do on the moon, and they weigh more on Jupiter than they do on Earth. This is because Jupiter has more mass than the Earth, which has more mass than the moon, and objects with greater mass have a stronger gravitational attraction than those with smaller mass. In SI units, kilograms are used, which is actually a measure of mass, not of weight; everything has the same mass, regardless of where it is located. However, when you want to apply mass to a physics problem, such has how long it will take an object to fall from a high-rise building, you have to include gravity (9.8 m/s^2 on Earth) in the solution.

Most liquids in U.S. supermarkets are measured in fluid ounces, which we have already discussed is a measure of volume. Sometimes, instead, items in the supermarket are measured in "net wt oz," which is "net weight ounces." Yes, there are two kinds of ounces. There are 16 net weight ounces in 1 pound. The weight of objects measured in fluid ounces depends on the density of the object. However, in SI units, most items in the supermarket are measured in grams, which is a measure of mass. If you are given a recipe for cookies that was created in Europe, the amounts of many ingredients will be written in grams (which is mass) rather than in cups.

The conversions for grams work the same way as they do for meters. Here, grams is the "base unit."

Prefix	Symbol	Equivalency	Number of Grams in Each
teragram	Tg	0.000,000,000,001	1,000,000,000,000
gigagram	Gg	0.000,000,001	1,000,000,000
megagram	Mg	0.000,001	1,000,000
kilogram	kg	0.001	1000
hectogram	hg	0.01	100
dekagram	dag	.1	10
gram	g	1 gram	1 gram
decigram	dg	10	0.1
centigram	cg	100	0.01
milligram	mg	1000	0.001
microgram	μg	1,000,000	0.000,001
nanogram	ng	1,000,000,000	0.000,000,001
picogram	pg	1,000,000,000,000	0.000,000,000,001

As with meters, use the column "Equivalency" to make conversions between the units. The units in this column are given numbers representing equal quantity. For instance, 0.001 kg is equal to 100 centigrams, which are both equal to 1 g. The numbers in the "Number of Grams in Each" column gives the number of grams that equal one such unit. For instance, there are 100 g in a hecktogram, and there are 0.001 g in a milligram. It is the "Number of Grams in Each" column that give the units their names. That is "kilo" means 1000, and so "kilogram" is 1000 g. The prefix "centi" represents 100th. There is 100th of a gram in a centigram; or you can think of it as 100 centigrams in a gram.

One kilogram is equal to 2.2046 pounds. You can use this equivalency to do conversions in the same way we did conversions in Section 8.2.

Understanding the Text:

1. When going on a camping trip in a remote location, such as the Boundary Waters in Minnesota or in the Rocky Mountains, you have to carry all of your food with you. It is recommended that you bring 2 pounds of food per day per person on your trip [2]. Since some things in the supermarket are measured in fluid ounces and some things are measured in net weight ounces, it is easier to measure the weight of things you buy for your trip in grams. If you need 2 pounds of food per day, how many grams of food per day per person do you need to bring?

2. Look around your house, or search online, for a combination of food you could pack for meals for one day (breakfast, lunch, dinner, and 3 snacks) that weighs 2 pounds

or less total, gives you a total of between 3500 and 4000 calories, and includes protein (meat or nuts), fruit, veggies, and fat. You will be able to use a fire or camp stove for breakfast and dinner, but not for the snacks or for lunch.

8.6 Other Measurements

In the United States, we tend to measure temperature in Fahrenheit. Almost everywhere else in the world, temperature is measured in Celsius. There is a linear relationship between the two different temperature scales, which we can find by finding the equation of a line through two equivalent temperature points. For instance, water freezes at 0° Celsius and boils at 100° Celsius. Water freezes at 32° Fahrenheit and boils at 212° Fahrenheit. To find a line through these two points, we use the general equation of a line in point-slope form:

$$m = \frac{y - y_0}{x - x_0} \tag{8.1}$$

What we have found above are two points on the graph of Celsius and Fahrenheit, where the x-values are in Fahrenheit and the y-values are in Celsius. The two points are $(32, 0)$ and $(212, 100)$.

The slope (m) is the difference in the y-values over the difference in the x-values, which many of you may have learned as "rise over run." Therefore, m in this case is:

$$\frac{100 - 0}{212 - 32} = \frac{100}{180} = \frac{5}{9}$$

Using the Equation 8.1 above, we use the slope we just found for m, and then use either of the two points we found for (x_0, y_0). I choose to use the point $(32, 0)$, because it has a zero in it and therefore looks simpler.

The equation of the line is then $\frac{5}{9} = \frac{y - 0}{x - 32}$. If we solve this for y, we get the relationship between Celsius and Fahrenheit in a form we are more used to seeing:

$$y = \frac{5}{9}(x - 32)$$

Most of the time, this is written using F for Fahrenheit and C for Celsius:

$$C = \frac{5}{9}(F - 32)$$

To use this equation, we put the number of degrees in Fahrenheit in for F in the equation and do the math on the right-hand-side of the equation– be sure to do what is in the parentheses first! The answer will be the temperature in Celsius.

If we are given degrees in Celsius and we want to change into Fahrenheit, we put the number of degrees in Celcius in for C and solve for F. We can solve for F first, if we want to. Then we get:

$$\frac{9}{5}C = F - 32$$
$$\frac{9}{5}C + 32 = F$$
or
$$F = \frac{9}{5}C + 32$$

As mentioned in the introduction to this chapter, to get a feel for temperature in Celsius, change the display in your car to Celsius. Eventually, you will recognize what temperatures are cold, mild, and hot in Celsius.

Although we in the United States are very attached to our system of measurements (feet, inches, pounds, etc.), there are many things that we are very comfortable with in the system of SI measurements. For instance, we use bytes, kilobytes, megabytes, and gigabytes, all of which use the SI prefixes; a kilobyte is approximately 1000 bytes, a megabyte is approximately 1000 kilobytes, and a gigabytes is approximately 1000 megabytes. Since computers do everything in base 2, there are actually 1024 bytes in a kilobyte, 1024 kilobytes in a megabyte, etc. The number 1024 is 2^{10}.

There are many more units of measurement that are not included in this chapter. However, this chapter should have given you a sense of how to convert between measurements, regardless of what the measurements are.

Understanding the Text:

1. You are on vacation in Rome, and you see that the weather forecast predicts the temperature will be $42°C$ by 11 am tomorrow. What temperature is that in Fahrenheit? How should you prepare for that weather?

2. You plan to go on vacation in Denmark in June. You find a website online that tells you the average temperature in Copenhagen, Denmark, in June is $19°C$ [7]. What is the temperature in Fahrenheit? What type of clothes should you pack for your trip? By the way, the average rainfall in Copenhagen in June is 49 mm [7]. Will that affect what you pack for your trip?

3. On the T (the Boston transit system) one day, you meet someone from Denmark who is vacationing in Boston. You warn him that it is going to be warm that day, but he wants to know exactly how warm it will be. You have no cell phone reception on the T, but luckily you recall the formula to convert from Fahrenheit to Celcius. You heard it was going to be $95°F$ in Boston that day. How many degrees is that in Celcius?

Chapter 9

Scaling

9.1 Scaling Objects on the Plane

In Section 6.4, we discussed isometries, which are movements that do not change the shape or orientation of an object. In this section, we talk about scaling, which changes all lengths of an object by the same factor. For example, in the each of the pictures below, the object to the left is the original object, and the object to the right (in blue) has been scaled by a factor of 3. That is, they are three times larger than the original object.

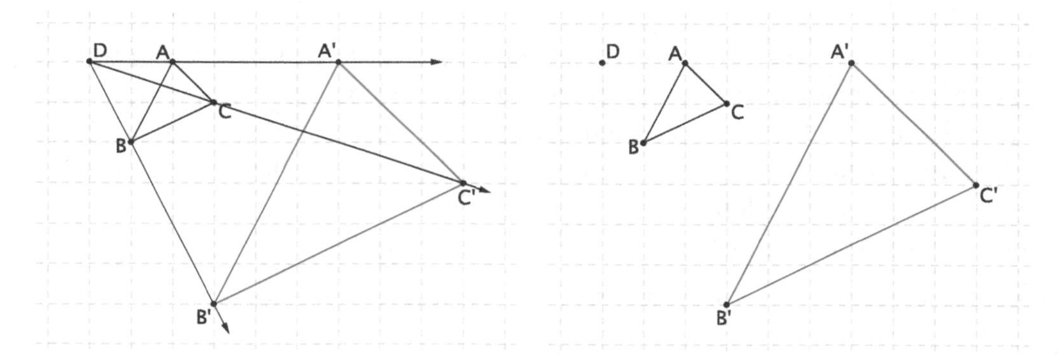

Not only are these triangles scaled by a factor of three, but their position is also special. The second (blue) triangle is three times as far from the point D (called the "center") as the original (black) triangle. In particular, look at the points A and A'. The point A is 2 units to the right of the point D. The point A', which is the transformed point corresponding to the point A, is 6 units to the right of D. Therefore, A' is three times as far from the point D as A is, and A' lies on the ray $\overrightarrow{DA}$.

Similarly, the point C' is three times farther, in the same direction, as C is from the point D. The point C is down 1 unit and to the right 3 units from Point D. The point C' is down 3 units and to the right 9 units from Point D. Note that both C and C' lie on the same ray originating from the point D. This is one way to check to see if you did the scaling correctly.

Here is another example. This example is of a square with scale factor $\frac{1}{2}$; the center of this scaling operation is Point O.

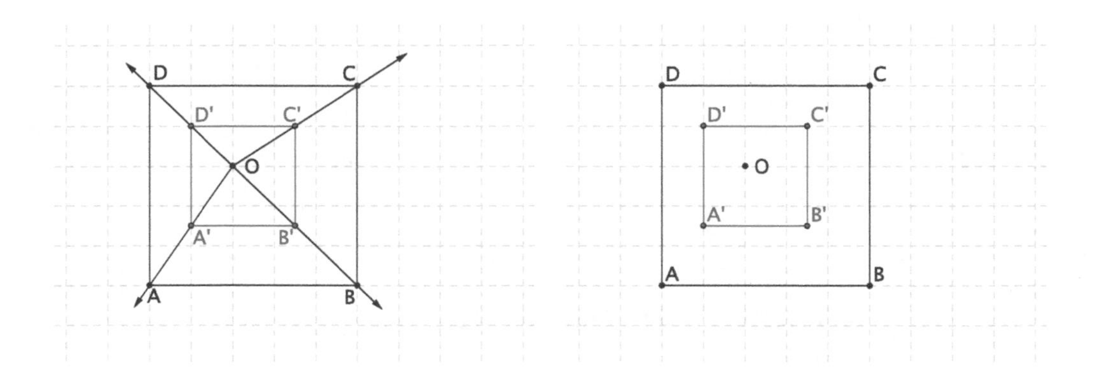

The original square in the above example is the outside square, which is drawn in black. Since the center of the scaling operation is Point O, every point of square $\square A'B'C'D'$ is half the distance from Point O as their corresponding points in $\square ABCD$ are from the point O. In fact, A' is on the ray $\overrightarrow{OA}$, B' is on the ray $\overrightarrow{OB}$, and so on. The resulting picture is the picture on the left.

How do you find these points yourself? Point D is 2 units to the left and 2 units up from the point O. Since we want the point D' to be half the distance between D and O, that means it should only be 1 unit to the left and 1 unit up from the point O. Similarly, the point A is 3 units down and 2 units to the left of O, so the scaled point, A' should be 1.5 units down and 1 unit to the left of O (i.e., half as far in both directions).

Now, let's work through another example. In the figure below, scale the figure by a factor of $\frac{1}{3}$ with center O. Try it on your own before continuing reading.

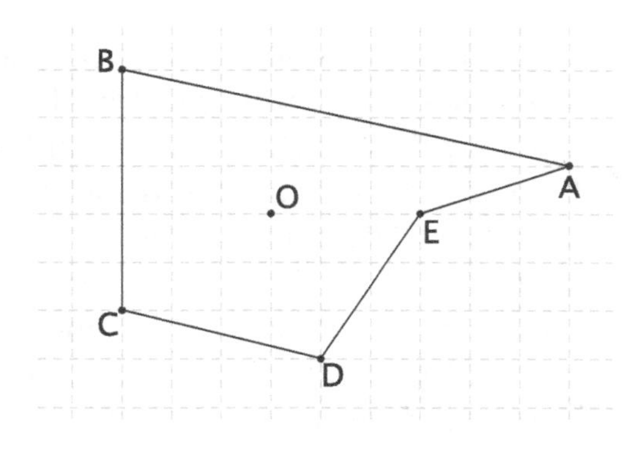

Since we have a scale factor of $\frac{1}{3}$, each point should be $\frac{1}{3}$ of the distance from O as the original point is; some of the points are not very complicated. The point B, for instance, is 3 units to the left and 3 units up from the point O. Since we are scaling by a factor of $\frac{1}{3}$, we multiply both of those distances by $\frac{1}{3}$ to find out where the scaled point should be. Since $\frac{1}{3} \times 3 = 1$, the scaled point, B' should be 1 unit to the left and 1 unit up from Point O, as in the picture below.

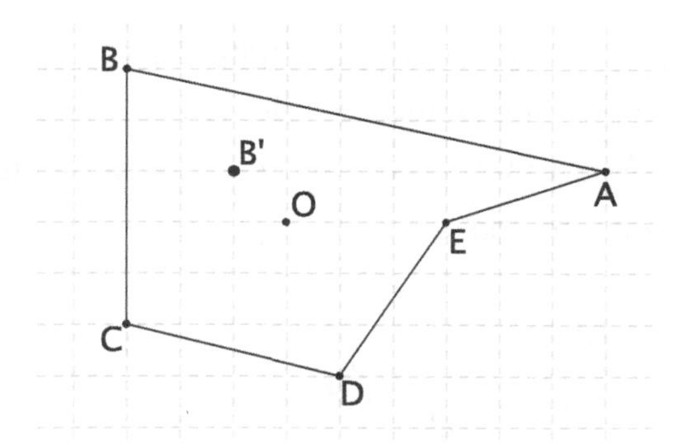

Point E is to the right 3 units from O, so E' should be to the right 1 unit from O because $3 \times \frac{1}{3} = 1$. Since E is on the same level with O (it is 0 units up from Point O), Point E' should be zero units up from O. That is, $0 \times \frac{1}{3} = 0$. The point E' is shown in the figure below.

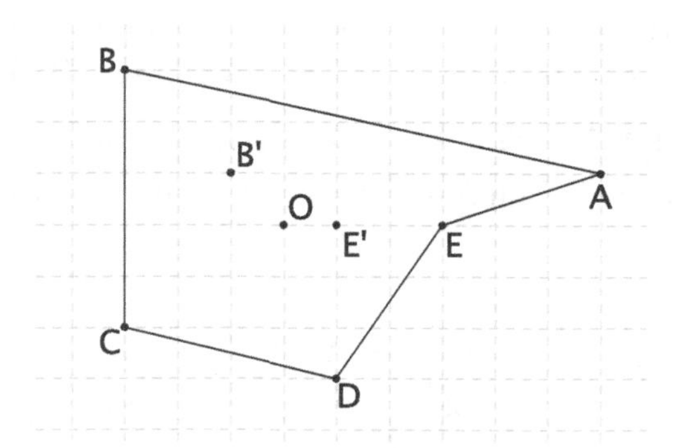

Now let us look at the point A. Point A is 6 units to the right of Point O and 1 unit up. Therefore, the scaled point, A' should be $6 \times \frac{1}{3} = 2$ units to the right of Point O and $1 \times \frac{1}{3} = \frac{1}{3}$ units up from Point O. The point A' is shown in the figure on the following page.

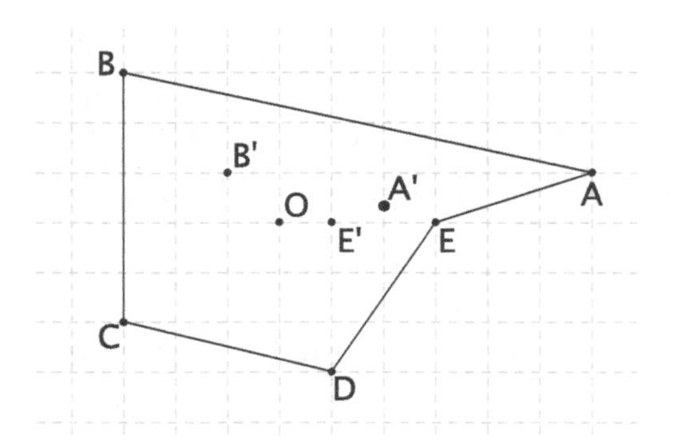

Point D' is similar to point A'; try it yourself before reading on.

The point D is 3 units down and 1 unit to the right of Point O. Therefore, Point D' should be $3 \times \frac{1}{3} = 1$ units down and $1 \times \frac{1}{3} = \frac{1}{3}$ units to the right of Point O. The picture below shows everything we have done so far.

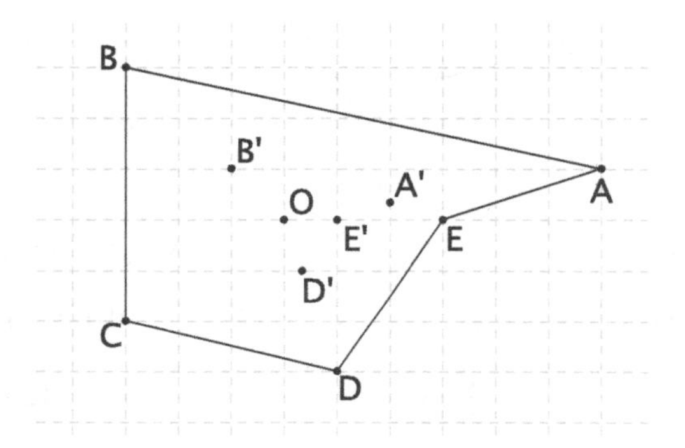

The only point remaining to scale is Point C. The point C is 3 units to the left and 2 units down from the point O. Therefore, the scaled point, C' should be 1 unit to the left and $\frac{2}{3}$ units down from Point O. Finally, connect the scaled points as the original points were connected to make the scaled shape. The final figure is shown below.

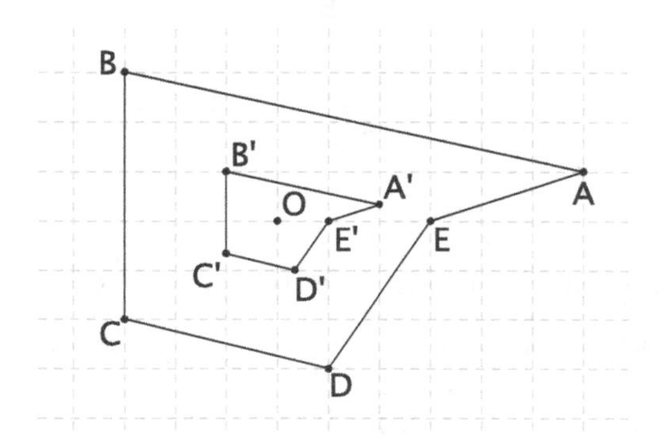

Understanding the Text:

1. Scale the figure below with center D and scale factor 3. Label the scaled points as in the examples in this section.

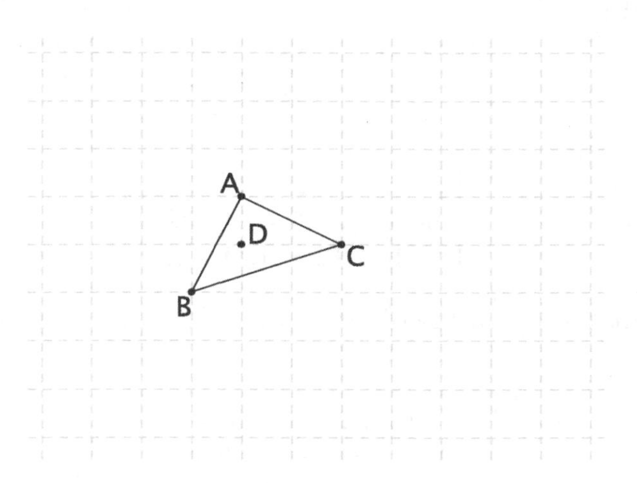

2. Scale the figure below with center E and scale factor $\frac{2}{3}$. Label the scaled points as in the examples in this section.

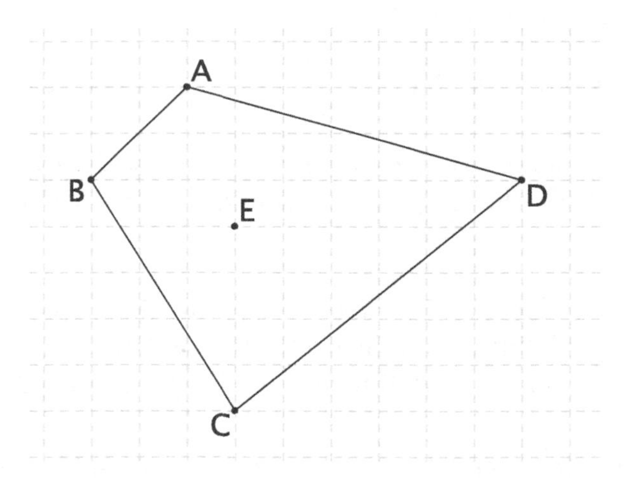

9.2 Scale Models

Sometimes the position of the object in space (or on a plane) is not important. Sometimes, all we want to do is create a model of a real-life object. Similarly, we may be interested in finding the real-life dimensions of a model of an object.

For instance, dollhouse furniture are models of furniture in real life. Imagine, for instance, that a dollhouse contains a coffee table that is 2 inches wide, 1.5 inches deep, and 1 inch high. How big would that coffee table be in real life? Before we can answer that question, we first need to know the scaling factor. That is, how much smaller is the real-life dollhouse furniture than a real-life object? In this case, the real life furniture is 12 times larger than the dollhouse furniture. This is denoted by the factor 1:12. Now we can answer the question:

The real-life furniture is:

$$\begin{gathered} 2 \text{ inches} \times 12 = 24 \text{ inches wide} \\ 1.5 \text{ inches} \times 12 = 18 \text{ inches deep} \\ 1 \text{ inch} \times 12 = 12 \text{ inches tall} \end{gathered}$$

If we then convert these real-life dimensions to feet, we see that the real-life coffee table is:

$$\begin{gathered} 24 \text{ inches} = 2 \text{ feet wide} \\ 18 \text{ inches} = 1.5 \text{ feet wide} \\ 12 \text{ inches} = 1 \text{ foot wide} \end{gathered}$$

Suppose we want to create a model given the dimensions of a real-life object. For example, say we want to create a model of a 2014 Toyota Tundra. According to the Toyota website [1], the dimensions of the real-life car are:

Height: 76 inches, Width: 79.9 inches, Length: 228.9 inches

In order to create a model, we need to decide what scaling factor we want to use. If we use a scaling factor of 1:24, the model truck would have the dimensions:

$$\begin{gathered} \text{Height: } 76 \text{ inches} \div 24 \approx 3.1667 \text{ inches} \\ \text{Width: } 79.9 \text{ inches} \div 24 \approx 3.3292 \text{ inches} \\ \text{Length: } 228.9 \text{ inches} \div 24 \approx 9.5375 \text{ inches} \end{gathered}$$

In this case, the model would be $\frac{1}{24}$ the size of a real-life truck. The symbol $\approx$ above means that the answer is approximate, rather than exact. This means that there is some rounding error in the answer.

Understanding the Text:

1. Suppose you would like to design a book for a doll house. The real-life book you would like to build a model of is 18 inches high, 12 inches wide, and 1 inch thick. Suppose you would like your model to be a 1:24 scale model. What would be the dimensions of the model book?

2. A $\frac{1}{25}$ model of a Slingster Dragster is $5\frac{15}{16}$ inches long, $2\frac{1}{2}$ inches wide, and $1\frac{3}{4}$ inches high [16]. What are the dimensions of the real-life car?

Chapter 10

Proportions and Geometric Sequences

10.1 Proportions

In order to find the proportion of a figure that is shaded (or unshaded), divide the figure into smaller figures of the same size and shape. Then add up the number that are shaded (or unshaded) and the total number. The proportion of the object that is shaded is:

$$\frac{\text{Number of Shaded Figures}}{\text{Total Number of Figures}}$$

For example, consider the square below. What proportion of the square is shaded?

Since the nine interior squares are the same size and shape, all we need to do is count squares. In this case, five squares are shaded out of a total of nine squares. Therefore, there $\frac{5}{9}$ of the square is shaded. This can also be written as a decimal, $0.555\overline{5}$, or as a percent, $55.5\overline{5}\%$.

Sometimes, when counting figures, the figures are not originally the same size. For instance, consider the square below:

The four completely unshaded squares are one size, but the shaded squares are smaller. In order to find the proportion of this square that is shaded, we first need to divide the larger squares into squares the same size and shape as the smallest squares, as shown in the figure below:

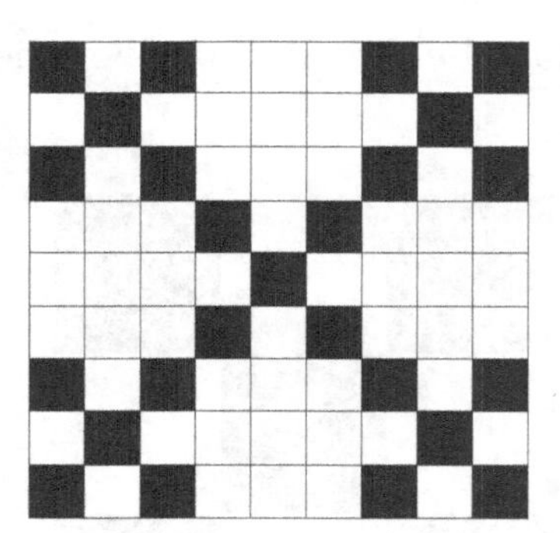

We can count each individual square, or we can use some mathematics to count the squares and make things faster for ourselves. To count the shaded squares, we can notice that there are five shaded squares in an "x" pattern in five locations in the figure. Therefore, there are $5 \times 5 = 25$ shaded squares in the figure. How many total squares are there? We can think of the figure as divided into nine squares, and then each of those nine squares are divided into nine smaller squares. Therefore, there are $9 \times 9 = 81$ total small squares in the figure.

This means that the proportion of the figure that is shaded is: $\frac{25}{81}$. This fraction cannot be reduced because $25 = 5 \times 5$ and $81 = 3 \times 3 \times 3 \times 3$, so the two numbers do not have any prime factors in common.

The proportion of the object that is unshaded is:

$$\frac{\text{Number of Unshaded Figures}}{\text{Total Number of Figures}}$$

However, there is another way to think about the number of unshaded figures. If the entire figure were shaded, then 100% of the figure would be shaded. As a proportion, this would be $\frac{\text{Total Number of Figures}}{\text{Total Number of Figures}} = 1$. Therefore, the number of unshaded figures can also be thought of as

$$1 - \frac{\text{Number of Shaded Figures}}{\text{Total Number of Figures}}.$$

In this case, since $\frac{25}{81}$ of the figure is shaded, the portion that is unshaded is $1 - \frac{25}{81} = \frac{56}{81}$.

When we look for patterns in the next section, it will be helpful to know this. Sometimes it is easier to find the proportion of shaded figures than the proportion of unshaded figures. It is helpful to know that given the proportion that is shaded, you can easily find the proportion that is unshaded. We will also find that it is helpful to leave the proportions written as fractions as we try to find a general formula for the proportion shaded at each stage.

10.2 Geometric Sequences

Geometric sequences are sequences resulting from iterative geometric patterns. For instance, the patterns used for the examples in the last section create a geometric sequence.

Stage 0

Stage 1

Stage 2

At the beginning (Stage 0), the entire object is shaded. At each successive stage, the shaded portion of the previous stage is divided into nine squares, and each set of nine squares is shaded in the same pattern. At Stage 1, the one shaded square from Stage 0 is divided into nine squares and then the four corner squares and the middle square are shaded. At Stage 2, the five shaded squares from Stage 1 are each divided into nine squares and for each set of nine squares, the four corner squares and the middle square are shaded. What would Stage 3 look like? First, redraw Stage 2 without any squares shaded. Then, break each square that was shaded in Stage 2 into nine squares by dividing each square into thirds in both directions. Finally, in each set of nine squares, shade in the corner squares and the middle square. The resulting figure should look like this:

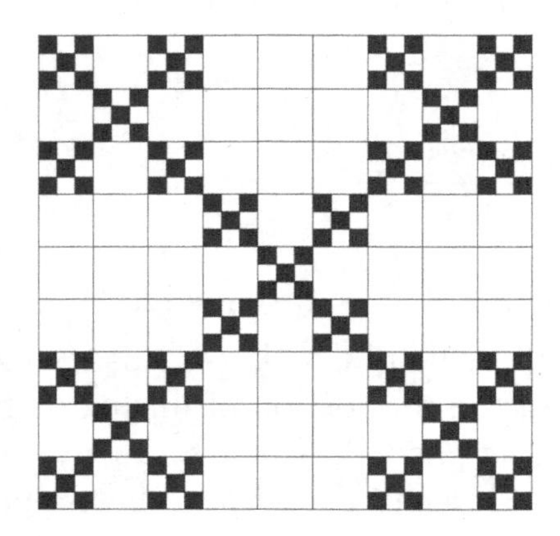

What proportion of this figure is shaded? First, for each of the shaded squares in Stage 2 there are five smaller shaded squares. Therefore, there are $25 \times 5 = 125$ shaded squares in Stage 3. Second, each of the 81 total squares in Stage 2 can contain nine squares that are the size of the shaded squares in Stage 3. Therefore, there are $81 \times 9 = 729$ total squares in the figure for Stage 3 that are the same size as the shaded squares in Stage 3. Finally, the proportion of the figure in Stage 3 that is shaded is $\frac{125}{729}$.

It is now rather unwieldy to actually create the figure for Stage 4. At this point, we hope we can find a pattern that will help us compute the proportion shaded at later stages.

To help us out with that, we create a table with columns for each stage and fill in the information we already have.

Stage	0	1	2	3	4	5	$\cdots$	50	$\cdots$	100	$\cdots$	n
Proportion Shaded	1	$\frac{5}{9}$	$\frac{25}{81}$	$\frac{125}{729}$			$\cdots$		$\cdots$		$\cdots$	

Next, we try to find the pattern. Perhaps we remember that at each stage, each shaded square from one stage became five shaded squares in the next stage. Similarly, each equally-sized square from one stage became nine squares in the next stage. Therefore, to get from one stage to the next, the numerator is multiplied by 5 and the denominator is multiplied by 9. Combining these two things, we notice that to get from one stage to the next, the proportion shaded in one stage is multiplied by $\frac{5}{9}$ to get to the next stage. Using that information in our table, we have the following:

Stage	0	1	2	3	4	5
Proportion Shaded	1	$\frac{5}{9}$	$\frac{25}{81}$	$\frac{125}{729}$		
Rewritten as Products	1	$\frac{5}{9}$	$\frac{5}{9} \times \frac{5}{9}$	$\frac{5}{9} \times \frac{5}{9} \times \frac{5}{9}$		

Repeated products can be rewritten as powers. For instance, $3 \times 3 \times 3$ can be written as 3^3. Therefore, we can rewrite the products in our table as:

Stage	0	1	2	3	4	5
Proportion Shaded	1	$\frac{5}{9}$	$\frac{25}{81}$	$\frac{125}{729}$		
Rewritten as Products	1	$\left(\frac{5}{9}\right)^1$	$\left(\frac{5}{9}\right)^2$	$\left(\frac{5}{9}\right)^3$		

Do you see a pattern? The next few entries in the table would be:

Stage	0	1	2	3	4	5
Proportion Shaded	1	$\frac{5}{9}$	$\frac{25}{81}$	$\frac{125}{729}$		
Rewritten as Products	1	$\left(\frac{5}{9}\right)^1$	$\left(\frac{5}{9}\right)^2$	$\left(\frac{5}{9}\right)^3$	$\left(\frac{5}{9}\right)^4$	$\left(\frac{5}{9}\right)^5$

We could fill in the numbers for the "Proportion Shaded" portion of the table as well, but our goal here is to find the pattern, so we will not.

Continuing with this pattern, which we are convinced works because of how the figures at each stage are created, we find the proportion of the figure that is shaded at the 50th stage:

$$\left(\frac{5}{9}\right)^{50}$$

At the 100th stage:

$$\left(\frac{5}{9}\right)^{100}$$

And finally, at the nth stage, for any stage n:

$$\left(\frac{5}{9}\right)^{n}$$

What if we want to know what proportion of the figure that is unshaded at each stage? As we discovered in Section 10.1, we can use the fact that the unshaded portion of the figure is the *rest* of the figure. Therefore, the proportion of the figure that is unshaded at each stage is:

Stage	0	1	2	3	4	5
Proportion Shaded	1	$1-\frac{5}{9}$	$1-\frac{25}{81}$	$1-\frac{125}{729}$		
Rewritten as Products	1	$1-\left(\frac{5}{9}\right)^1$	$1-\left(\frac{5}{9}\right)^2$	$1-\left(\frac{5}{9}\right)^3$	$1-\left(\frac{5}{9}\right)^4$	$1-\left(\frac{5}{9}\right)^5$

Continuing with this pattern, we find the proportion of the figure that is unshaded at the 50th stage:

$$1-\left(\frac{5}{9}\right)^{50}$$

At the 100th stage:

$$1-\left(\frac{5}{9}\right)^{100}$$

And finally, at the nth stage for any stage n:

$$1-\left(\frac{5}{9}\right)^{n}$$

Appendix A

Constructions

A.1 Constructing a Line Parallel to Another Line Through a Given Point

Given a line and a point not on that line, we will now see how to construct a line through the point that is parallel to the given line using only a compass and a straight edge (a straight edge is like a ruler where you do not use the marks for measurements).

a. On your paper, draw a line and a point that is not on that line. These will be your given point and your given line. It will help to have at least $\frac{1}{2}$ inch between the point and the line in order to get an accurate picture, but any line and point will work. Call the point P and the line l.

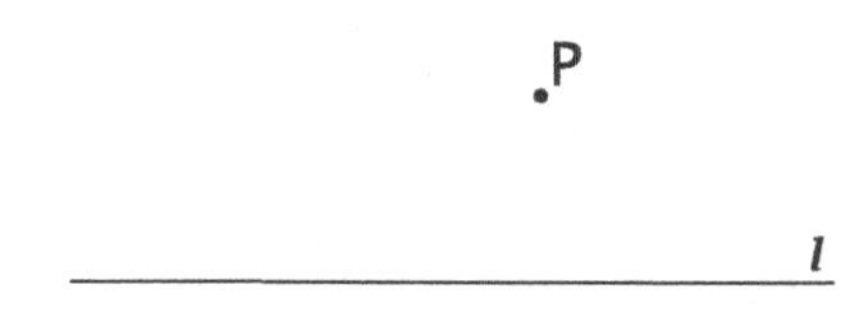

b. Open your compass wide enough so that if you put the point of your compass on the point you drew in Step A.1a., the compass will cross (intersect) the line you drew. Draw a short arc through one such point of intersection, and call that intersection point A.

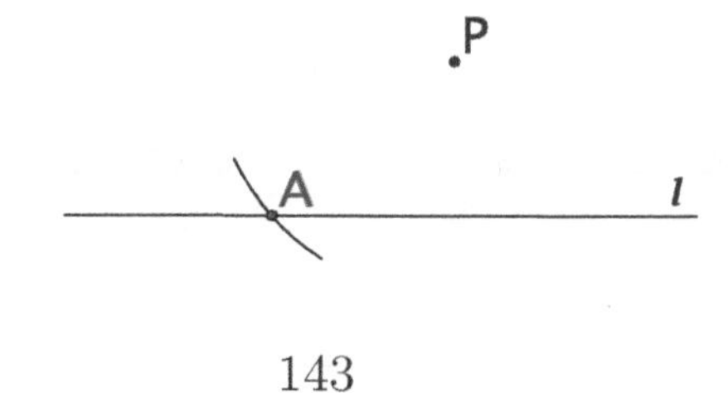

c. Without changing the opening of your compass, put the point of your compass on Point A and draw another arc of intersection with line l. Call that point B. If you had already changed your compass opening, go back and make it the same size at the distance between Point P and Point A.

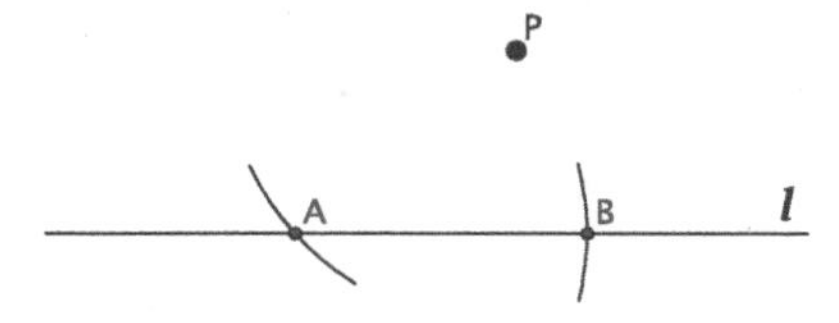

d. Again without changing the opening of your compass, put the point of your compass on Point B and draw an arc on the same side of P as Point B. This arc should be directly to the right (or left) of Point P when your given line is horizontal on the page.

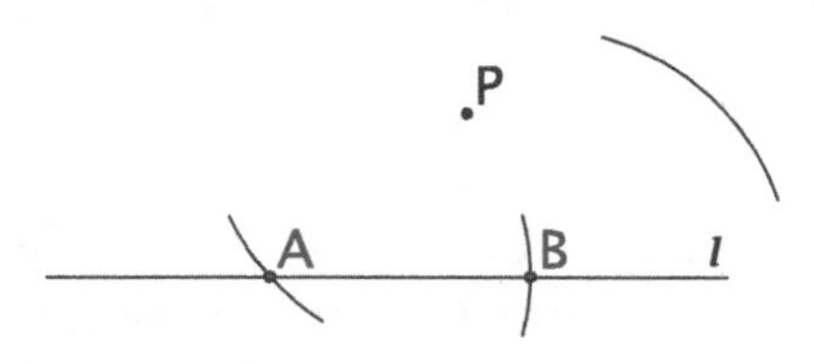

e. Next, without changing the opening of your compass, put the point of your compass on Point P and draw an arc that crosses the arc you drew in Step A.1d.

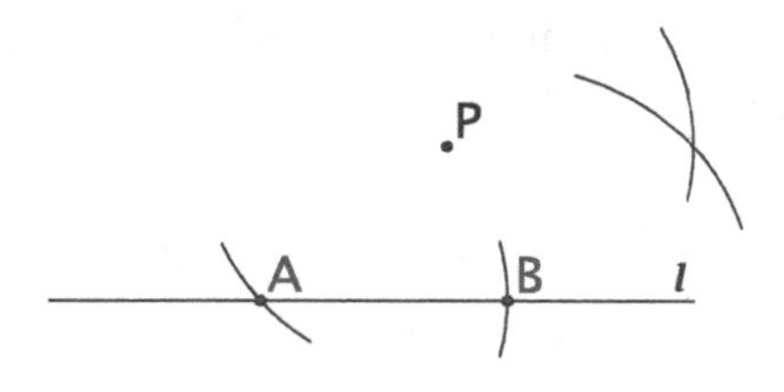

f. Draw a line through the intersection from Step A.1e. and Point P. This line should be parallel to Line l, and it goes through the Point P.

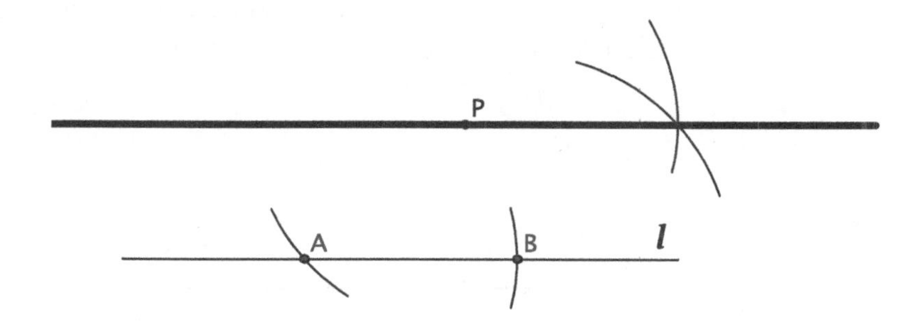

A.2 Constructing a Line Perpendicular to a Given Line

a. Begin with a line on a piece of paper. This is your "given line."

b. Create two points on your line and label them A and B. For the accuracy of your drawing, these points should be at least 1 inch apart.

c. Put the point of your compass on A, the pencil of your compass on B, and draw a circle with radius $\overline{AB}$.

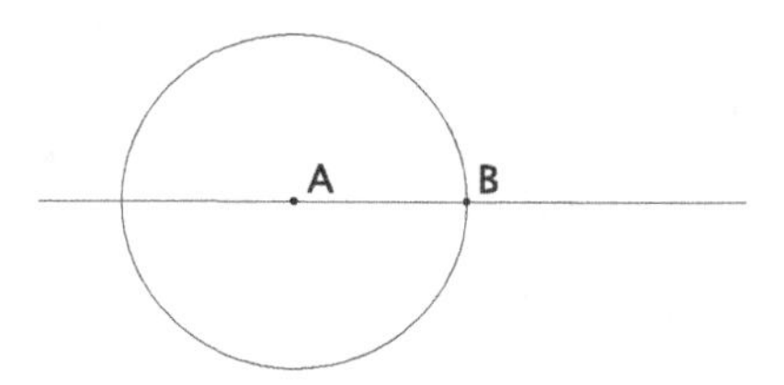

d. Put the point of your compass on B, the pencil of your compass on A, and draw a circle with radius $\overline{AB}$.

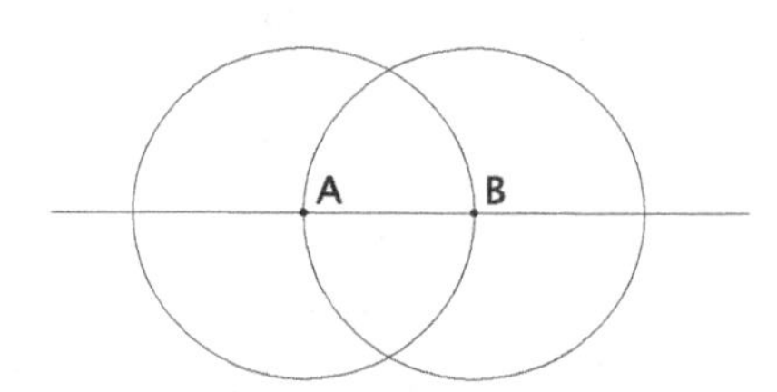

e. Draw a line through the two places the circles you made intersect. This line is perpendicular to the given line.

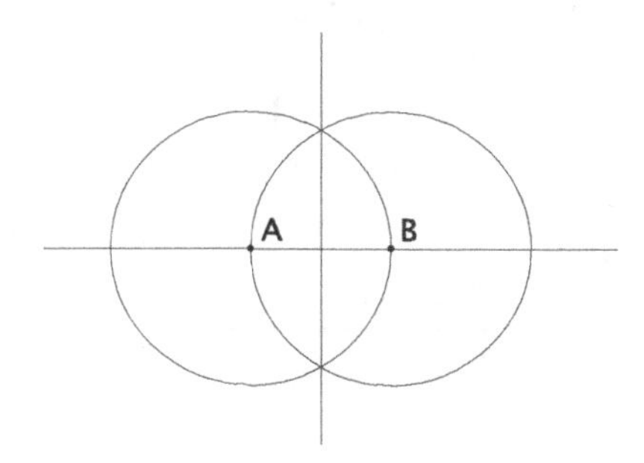

A.3 Constructing a Line Perpendicular to a Given Line Through a Given Point

a. Begin with a line on a piece of paper and a point that is not on that line. This is your "given line" and your "given point." For the accuracy of your drawing, it helps if your point is at least 2 finger widths away from your line.

b. Label the arbitrary point P.

c. Put the center (or "point") of your compass on the point P and use your compass to draw an arc that crosses the given line. Call that point A.

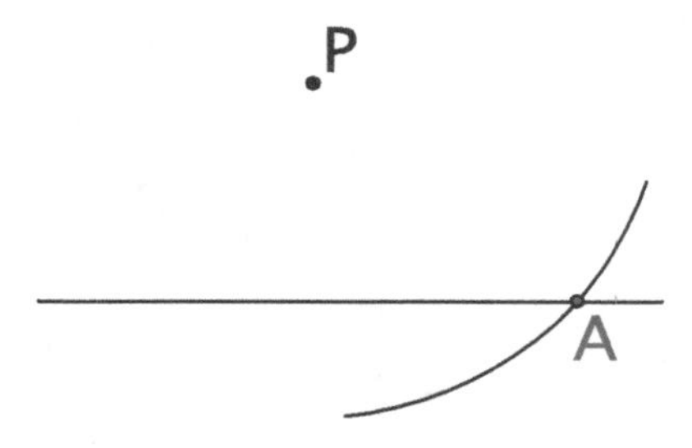

d. Leaving your compass open the same amount as in Step A.3c., keep the center (or "point") of your compass on the point P and use your compass to draw a second arc that crosses the given line. Call that point B.

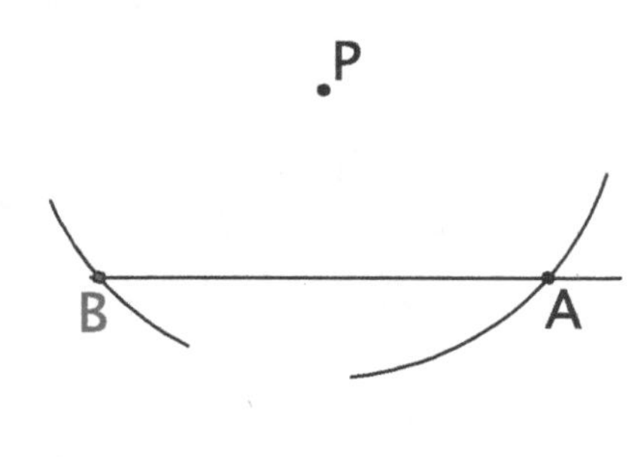

e. Put the center of your compass on Point A and draw an arc on the opposite side of $\overleftrightarrow{AB}$ from Point P.

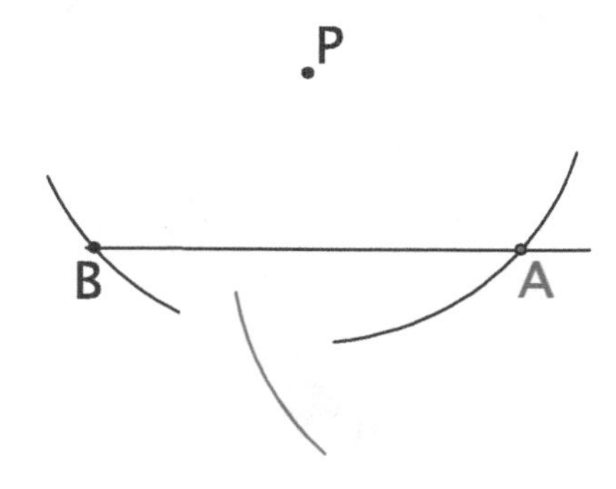

f. Leave the opening of your compass the same as in Step A.3e., put the center of your compass on Point B, and draw an arc that intersects the arc you made in Step A.3e.. Call the intersection point Q.

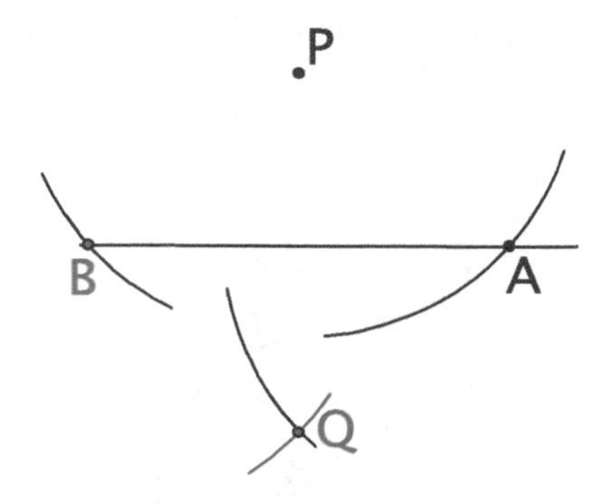

g. The line $\overleftrightarrow{PQ}$ is the line perpendicular to the line AB through the given point P.

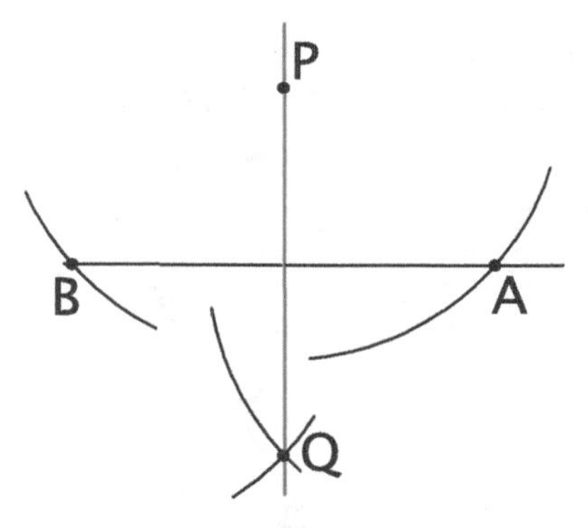

You can see an animation of this construction, as well as a proof that it works at `http://www.mathopenref.com/constperpextpoint.html`.[13]

A.4 Bisecting an Angle

a. Begin with an angle on a piece of paper. This is your "given" angle. Call the vertex of your angle Point A.

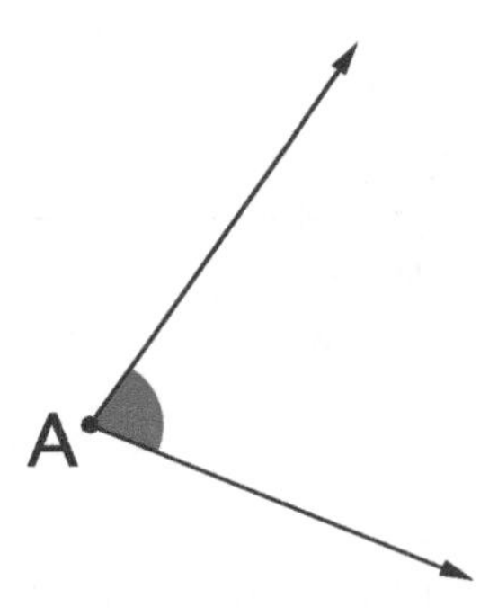

b. Put the center of your compass on Point A and draw an arc that intersects both of the rays of your angle. Call the intersection points B and C. For this step, your compass may have any radius, but for practical reasons, make the radius large enough so you can make an accurate arc, and not so large that the compass is almost completely open.

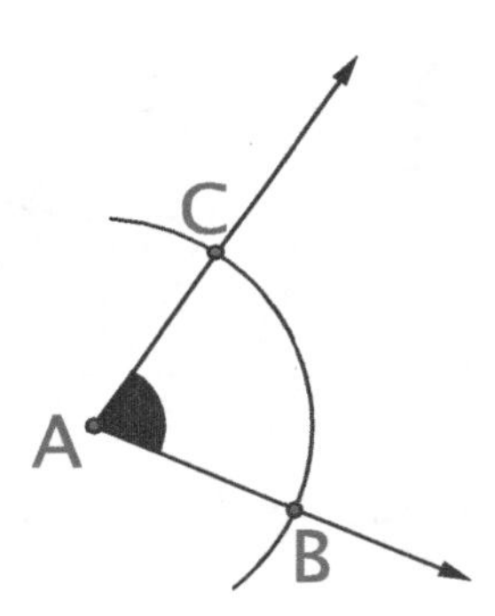

c. Put the point of your compass on Point B and draw an arc roughly in the center of your angle, "outside" of the arc you drew in Step A.4b.. Here, again, it does not matter how open your compass is.

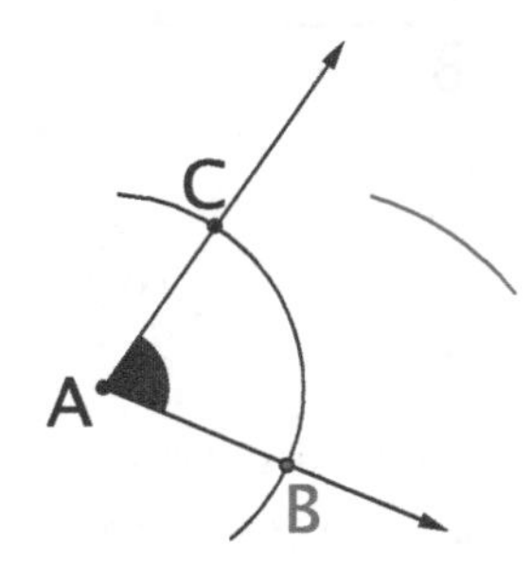

d. Leave the opening of your compass the same as it was for Step A.4c. above and put the point of your compass on point C. Draw and arc that crosses the arc you made in Step A.4c. Label the intersection point D.

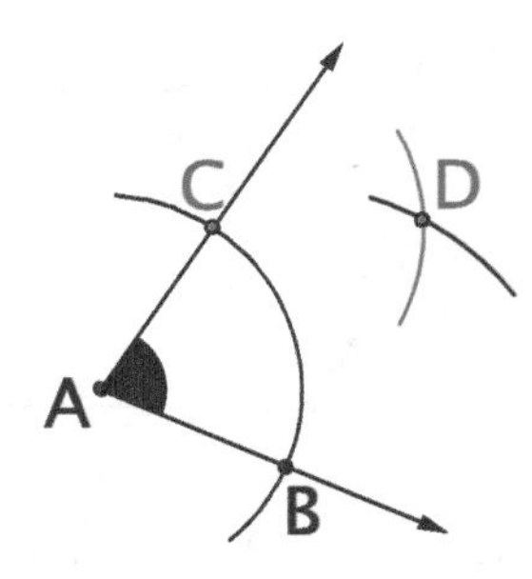

e. Draw the ray $\overrightarrow{AD}$. Now the angles $\angle BAD$ and $\angle CAD$ are half the size of angle $\angle BAC$, and the angle bisector of $\angle BAC$ is $\overrightarrow{AD}$.

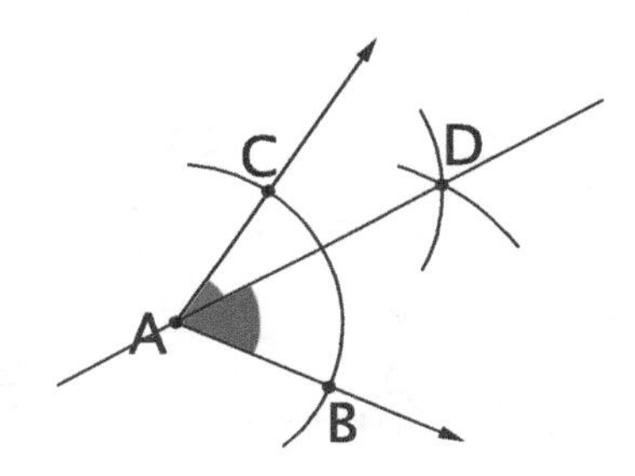

A.5 Constructing an Equilateral Triangle

a. Begin with a line segment of the length you want the sides of your triangle to be. Label your line segment $\overline{AB}$.

b. Put the point of your compass on the point A and the pencil of your compass on the point B. This is the compass radius you will keep for the remainder of this construction.

c. With the point of your compass on Point A and the same compass opening as in Step A.5b., draw an arc above and near the center of your line segment.

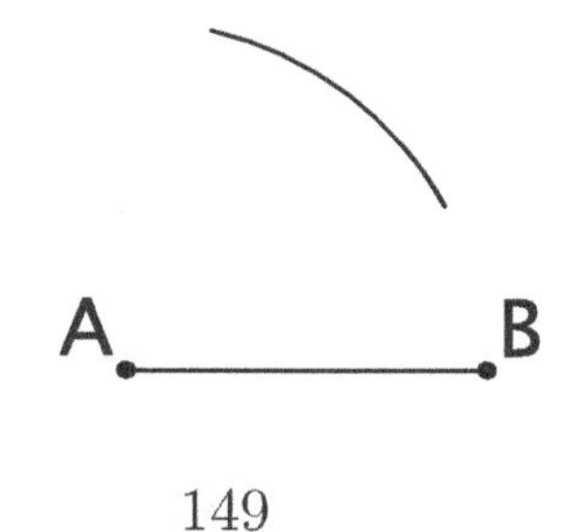

d. With the point of your compass on Point B and the same compass opening as in Step A.5b., draw an arc above and near the center of your line segment, intersecting the arc you created in Step A.5c..

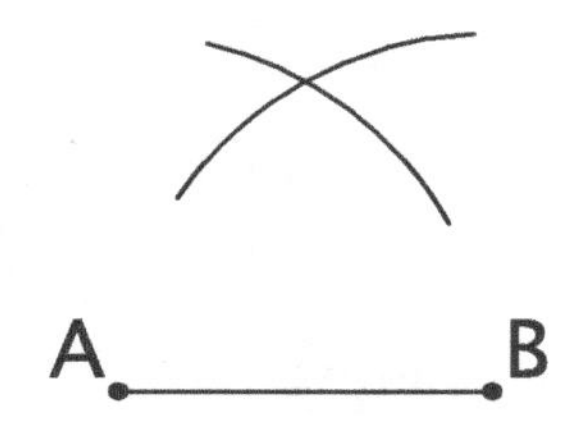

If the arcs in Step A.5c. and Step A.5d. cannot intersect, re-do Step A.5c. and Step A.5d., extending your arcs slightly farther until they cross each other.

e. Label the point of intersection of your two arcs C.

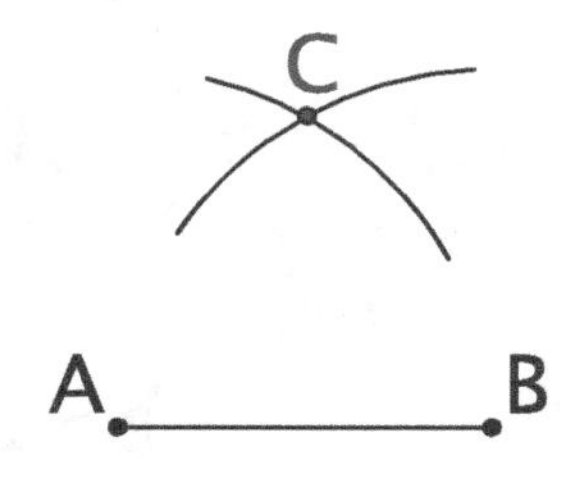

f. Draw the equilateral triangle ΔABC.

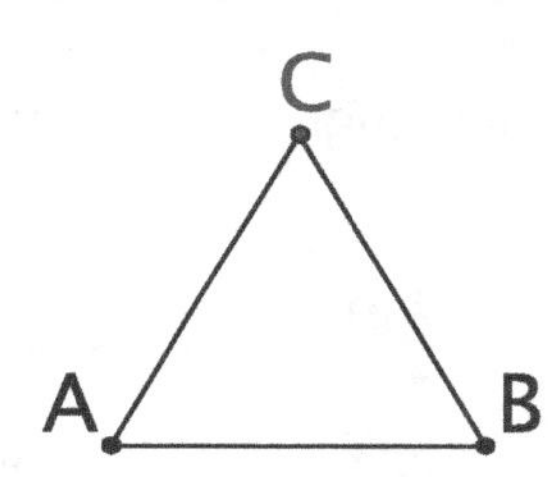

Note: If your triangle does not look equilateral, this probably means that you either skipped Step A.5b. or your compass opening moved in the process of carrying out the instructions.

Bibliography

[1] *2014 Toyota tundra features & specs*, December 2013, available at http://www.toyota.com/tundra/features.html#\protect\kern-.1667em\relax/dimensions/8240/8252/8275/8276.

[2] *3 Day Backpacking Menu–Section Hiker*, June 2013, available at http://sectionhiker.com/3-day-backpacking-menu/.

[3] *Area of a disk – Wikipedia, the free encyclopedia*, June 2013, available at http://en.wikipedia.org/wiki/Circle_area.

[4] *Buttercup*, June 2013, Photo by Amy Wangsness Wehe.

[5] *CIA– the world factbook, appendix g :: Weights and measures*, March 2013, available at https://www.cia.gov/library/publications/the-world-factbook/appendix/appendix-g.html.

[6] *Convex Polyhedron*, December 2013, Photo by Amy Wangsness Wehe.

[7] *Copenhagen, Denmark weather averages– monthly average high and low temperatures– average precipitation and rainfall days– world weather online*, June 2013, available at http://www.worldweatheronline.com/Copenhagen-weather-averages/Hovedstaden/DK.aspx.

[8] *Cube*, December 2013, Photo by Amy Wangsness Wehe.

[9] *Cylinder*, June 2013, Photo by Amy Wangsness Wehe.

[10] *Day Lily*, June 2013, Photo by Amy Wangsness Wehe.

[11] *Dodecahedron*, December 2013, Photo by Amy Wangsness Wehe.

[12] *Honeysuckle*, June 2013, Photo by Amy Wangsness Wehe.

[13] *How to construct (draw) a perpendicular from a line through a given point*, June 2013, available at http://www.mathopenref.com/constperpextpoint.html.

[14] *Icosahedron*, December 2013, Photo by Amy Wangsness Wehe.

[15] *Immobilien, wohnungen und häuser bei immobilien scout24 mieten, kaufen, inserieren*, December 2013, available at http://www.immobilienscout24.de/.

[16] *Monogram 1/25 Slingster Dragster Plastic Model Kit*, December 2013, available at http://www.revell.com/model-kits/cars/85-4997.html#.UsB-3fZvSLg.

[17] *NASA's metric confusion caused Mars orbiter loss– September 30, 1999*, March 2013, available at http://www.cnn.com/TECH/space/9909/30/mars.metric/.

[18] *Nonconvex Bridge*, December 2013, Photo by Amy Wangsness Wehe, figure created by Ben Cote.

[19] *Norwegian Flag*, June 2013, Photo by Amy Wangsness Wehe.

[20] *Purple Flower*, June 2013, Photo by Amy Wangsness Wehe.

[21] *Square Prism*, June 2013, Photo by Amy Wangsness Wehe.

[22] *Starpolyhedron*, December 2013, Photo by Amy Wangsness Wehe, figure created by Judy Hanson.

[23] *Triangular Prism*, December 2013, Photo by Amy Wangsness Wehe.

[24] *Triangular Prism for volume*, June 2013, Photo by Amy Wangsness Wehe.

[25] *Oxford Dictionaries*, August 2014, available at http://www.oxforddictionaries.com/.

[26] A.V. Astin, H. A. Karo, and F. H. Mueller, *Refinement of values for the yard and the pound*, Doc 59-5442, Federal Register, June 25, 1959, available at http://www.ngs.noaa.gov/PUBS_LIB/FedRegister/FRdoc59-5442.pdf.

[27] John Ayto, *Word origins*, 2 ed., A& C Black, London, October 2005.

[28] Robert P. Crease, *The world in the balance: The historic quest for an absolute system of measurement*, W. W. Norton & Company, Inc., New York, 2011.

[29] Victor J. Katz, *A history of mathematics*, Pearson, New York City, New York, 2008.

[30] Merriam-Webster, *Tessellation – definition and more from the free merriam-webster dictionary*, February 2013, available at http://www.merriam-webster.com/dictionary/tessellation.

[31] YouTube.com, *Measuring Angles With a Protractor - YouTube*, November 2012, available at http://www.youtube.com/watch?v=50-9wgGufvc.

Index